Ali Zuashkiani

Expert Knowledge Based Reliability Models: Theory and Case Study

Blair G

Ali Zuashkiani

Expert Knowledge Based Reliability Models: Theory and Case Study

Integrating Data and Expert Opinion Using Bayesian Statistics to Build Complex Reliability Models

VDM Verlag Dr. Müller

Imprint

Bibliographic information by the German National Library: The German National Library lists this publication at the German National Bibliography; detailed bibliographic information is available on the Internet at http://dnb.d-nb.de.

Cover image: www.purestockx.com

Publisher:
VDM Verlag Dr. Müller Aktiengesellschaft & Co. KG, Dudweiler Landstr. 125 a, 66123 Saarbrücken, Germany,
Phone +49 681 9100-698, Fax +49 681 9100-988,
Email: info@vdm-verlag.de

Produced in USA and UK by:
Lightning Source Inc., La Vergne, Tennessee, USA
Lightning Source UK Ltd., Milton Keynes, UK
BookSurge LLC, 5341 Dorchester Road, Suite 16, North Charleston, SC 29418, USA

ISBN: 978-3-639-02056-4

TABLE OF CONTENTS

LISTS OF FIGURES AND TABLES

Figures

Tables

Common Notations and Abbreviations Used in Chapters 7–13

PHM	Proportional hazard models
MCMC	Markov Chain Monte Carlo
CBM	Condition-based maintenance
RUL	Remaining useful life
ME	Mean estimator
MATOR	Machine simulator
CDND	Canadian Department of National Defence
CMMS	Computerized maintenance management systems
ML	Maximum likelihood
DOE	Design of experiments
C_f	Cost of a failure replacement
C_p	Cost of a preventive replacement
L_{Wi}	Warning limit of $Z_i(t)$
L_{ci}	Critical limit of $Z_i(t)$
T	A variable describing the time to failure
n	Number of samples
Q	Total number of blocks
w_p	The weight given to the pth expert
t_i	Current observation time
δ_i	A dummy variable which denotes the current state of the history
γ_i	Coefficient of ith time-dependent covariate
β	Shape parameter of Weibull distribution
η	Location parameter of Weibull distribution
A	$\ln(\eta^{\beta} / \beta)$
$Z_i(t)$	i^{th} time-dependent covariate
$Z_1(t_i), Z_2(t_i),...Z_m(t_i)$	Time-dependent covariates at t_i
$\beta^*_i, A^*_j, \gamma^*_{k1},..., \gamma^*_{1m}$	The mid points in block $b_{i,j,k,..l}$
E	Expert knowledge
D	Statistical data
D10	First 10 histories of statistical data
h_{E+Dj}	Hazard rate based on the expert knowledge combined with j^{th} data history
h_{D39}	Hazard rate based on MLE of 39 data histories
P_l	Lower bound for the probability of failure in a short interval $[t_0, t_0 + \Delta t_0]$
P_u	Upper bound for the probability of failure in a short interval $[t_0, t_0 + \Delta t_0]$
$f^{(n)}{}_{\beta, A, \gamma_1,...,\gamma_m}$	Empirical prior distribution of the PHM parameters
$\hat{g}_\theta(\beta, A,..., \gamma_m / y)$	Empirical posterior distribution of the PHM parameters

To Sara

1. INTRODUCTION

In recent years, the increasing need for competitiveness and therefore productivity has forced industries to delegate more manual tasks to industrial machinery, resulting in more automation. As a result, there is a greater need for keeping the machines working, which has attracted industry and academia alike to develop and apply more accurate reliability and maintenance techniques. This effort has resulted in the creation of condition-based maintenance (CBM), a recently developed field of maintenance capable of making use of all data related to a system for predicting the system's health conditions.

CBM not only measures the effect of the age of a system, but also takes into account other influential factors on the lifetime of a system, such as metal particles in the engine-circulating oil, vibration intensity, and temperature. To date, two major categories of techniques have been applied in CBM (Jardine, 2002; Zuashkiani et al., 2004):

1. Non-statistical tools including neural networks, fuzzy logic, and expert systems
2. Statistical tools including proportional hazards, log linear regression, linear regression, and proportional intensities models

All these techniques have their own advantages and disadvantages. However, statistical methods are among the most widely used tools in CBM and have been applied in this field for many years. Among them the proportional hazards model (PHM) is unique because its structure allows covariates or explanatory variables to affect the hazard rate of a system. This is a more sensible, real-world assumption than other tools. Another advantage of PHM that has made it more practical is its ability to handle time-dependent covariates. Moreover, unlike most conventional models, PHM is capable of handling right censored and tied survival data.

These features of PHM have caused some researchers to use PHM more widely in industrial applications (Banjevic et al., 2001; Jardine et al., 1987). The benefits and applicability of PHM led to the creation of EXAKT (Banjevic et al., 2001), a software package that uses PHM in estimating the hazard function of a system as a function of its condition indicators. In addition, EXAKT uses Markov chain modeling to describe the behavior of the condition over time. Afterwards, it considers cost of failure and cost of preventive replacement to find the optimum replacement time. EXAKT has been applied successfully in a few companies during the last decade (see section 11.3.1). However, a practical difficulty in developing statistical models including PHM arises from their need for large sets of data in proper formats in order to produce reliable outputs. At the same time, since implementing CBM is often more expensive than time based models, it is generally suggested that it be utilized only for critical equipment. On the other hand, critical equipment is well maintained which means there are usually a small number of failure data, or no failure data at all. To solve this problem, researchers have been seeking methodologies that can make use of other types of information such as experts' knowledge. This search resulted in employing Bayesian statistics. The structure of Bayesian methods allows incorporation of both experts' knowledge and statistical data in model building which is a necessity in many fields of study, particularly reliability.

However, there are two major obstacles to using Bayesian methods when incorporating experts' knowledge. First, it is difficult both to elicit such expert knowledge (knowledge elicitation) and to formulate that knowledge in a proper probabilistic format (knowledge formulation). The task becomes even more difficult when dealing with complex multi-parameter models which is the case when we build a CBM model.

Second, even with the use of Markov Chain Monte Carlo methods and current advances in computational facilities, the Bayesian updating process and sampling from the posterior distribution can be very complicated and time consuming for complex multi-parameter models. This is much worse when it comes to PHM with time-independent covariates and informative prior distributions. As a result we know of no published work on extracting knowledge, transferring knowledge to prior distribution, or updating prior to posterior distributions for PHM with parametric baseline hazard, time-dependent covariates, and informative prior distributions.

The research reported in this book has resulted in a methodology that can incorporate experts' knowledge as well as statistical data in estimating the parameters of a PHM with time-dependent covariates. It also presents a numeric updating process that can be used to update the values of parameters when new statistical data arrive. Furthermore, the results of this research are believed to be beneficial to other fields of study that currently use PHM such as biomedical, finance, and organization demography.

In the next section, the importance of reliability and maintenance, and especially the role of condition-based maintenance, is discussed. Different tools and techniques applied in CBM and some of their applications and advantages are then summarized. In Chapter 3 the focus is on the PHM, its definition, and its applications especially in reliability and maintenance. Chapter 4 examines the role of Bayesian statistics in solving today's practical problems in the areas of reliability and maintenance. Also reviewed are researchers' attempts to calculate the posterior distribution for PHM..

Chapter 5 provides a brief literature on different knowledge elicitation techniques that have been applied in practice over the last six decades., with emphasis on techniques relevant to this book. To distinguish between knowledge elicitation and knowledge formulation we present a review of knowledge formulation techniques in Chapter 6. The two concepts are inter-related and can be confused as the latter is sometimes referred to as *knowledge elicitation* by researchers and practitioners whose job is to formulate expert knowledge in a probabilistic format. Since problem definition is not introduced in detail until Chapter 7, we briefly define it here so that readers can relate the literature and background presented in the first six chapters to the problem:

The proportional hazards model with Weibull baseline hazard function and time-dependent covariates is considered to describe the hazard rate of a system (see the formula in the next page). The aim is to estimate the parameters of this model ($\beta, \eta,\ \gamma = (\gamma_1, \gamma_2 ..., \gamma_m)$) using both expert knowledge and data by applying knowledge elicitation techniques and Bayesian statistics (for more explanations and details see Chapter 7).

$$h(t; Z(t)) = \frac{\beta}{\eta}\left(\frac{t}{\eta}\right)^{\beta-1} e^{\gamma_1 Z_1(t) + ... + \gamma_m Z_m(t)}$$

A more detailed statement of the problem, followed by the results obtained in this research is presented in chapters 7, 8, 9, and 10. Chapter 11 discusses the results obtained during simulation experiments and a real world case study undertaken at Dofasco Inc. in 2003–2006. These results show how the techniques suggested in this research can be applied in industry. The conclusion of the research and suggested guidelines for future research in this area is the topic of the last chapter.

2. CONDITION-BASED MAINTENANCE

The area of reliability and maintenance was established in the early 1960's by researchers such as Barlow, Proschan, Jorgenson, McCall, Rander and Hunter. The results of their studies are summarized in a review by McCall (1965) and a book by Barlow and Proschan (1965). Wireman (1990) carried out a benchmarking exercise and found that the maintenance cost for industrial firms in the USA has grown by 10-15% per year since 1979. This highlighted the importance of developing more accurate reliability and maintenance tools. Models that use all related data to predict the failure time of a system attracted most of the attention. Such models generally fall under the Condition-Based Maintenance (CBM) category. A survey carried out by several major organizations has revealed that with an investment of US$10K–US$20K in condition monitoring, one can save up to USD $500,000 a year (Rao, B. K. N., 1996; Sasranga & Knezevic, 2001).

As is clear from its name, Condition-Based Maintenance is a branch of maintenance that works based on the condition of a system. More precise definitions of CBM are provided here to demonstrate how the concept has been clarified over time.

According to Bunks and McCarthy (2000), the term Condition-Based Maintenance is used to indicate the monitoring of machines for the purpose of diagnostics and prognostics. Diagnostics are used to determine the current health condition of a machine's internal components, prognostics are used to predict their remaining useful life.

Tse and Atherton (1999) define CBM as a technique used to reduce the uncertainty of maintenance actions, based on the needs indicated by equipment's condition. Condition-Based (predictive) Maintenance involves the irregular or constant collection and interpretation of data related to the operating condition of critical components of equipment, prediction of the incidence of failure, and determination of appropriate maintenance strategies and actions (Knapp et al., 2000).

It seems that CBM is designed to use all possible data related to the health condition of equipment in order to predict its future behavior, and based on that prediction, to make necessary decisions in advance. What is meant by equipment condition? How can it be measured?

Equipment condition can be estimated via some condition monitoring variables called *explanatory variables* or *covariates*. These are variables that are thought to represent the health condition of equipment. Some possible explanatory variables in industry are: continuous variables such as stress, temperature, voltage, pressure, vibration, acoustic emission, and level of metallic particles in the machine lubricating oil; discrete variables such as the number of hardening treatments or of simultaneous users of a system; and categorical variables such as manufacturer, design, and location (Meeker & Escobar, 1998).

A recently developed study in the area of maintenance indicates that there are three major problems facing many modern engineering plants (Tse & Atherton, 1999):

1- How maintenance actions for sophisticated equipment in a complex operating environment can be planned and scheduled ahead of time.

2- How inventory costs of spare parts can be reduced.

3- How the risk of catastrophic failures can be reduced, and how unplanned forced outage of equipment or systems can be eliminated.

All of the above can be addressed by developing and applying an effective CBM system. This requires models that can indicate when to do maintenance based on a system's condition.

As mentioned in the introduction, two major categories of techniques have been applied in CBM to address this issue:

1. Non-statistical tools including neural networks and expert systems
2. Statistical tools including *trending* or statistical process control (SPC), log linear regression model, linear regression model, and proportional hazards model (PHM)

Here we briefly describe each tool with its applications in CBM and some of its main advantages and disadvantages.

Note: although many categorize Neural Networks as a statistical tool, since its properties are different from conventional statistical tools some authors do not include it in the statistical models (Jardine, 2002).

2.1 Non-Statistical Tools in Condition-Based Maintenance

The advent of non-statistical tools goes back less than two decades. Non-statistical tools are techniques that are borrowed from outside of the statistical realm such as artificial intelligence (AI) and neural networks (NN). In the next sections basic definitions and background about expert systems and neural networks are presented including their advantages and disadvantages, and their application.

2.1.1 Expert Systems

For many years there has been interest in developing methodologies similar to those used by humans to deal with complex problems. Research done in this area is known as Artificial Intelligence (AI). One of the consequences of AI research has been the development of techniques capable of modeling information at high levels of abstraction. Based on these techniques some languages have been developed that can be used to create computer programs that closely resemble human logic in their implementation. Such programs, which copy human expertise in well-defined problem domains, are called *expert systems* (Riley, 2004).

Some definitions for expert systems are function-based, others are structure-based, and some consider both functional and structural aspects. Most early definitions assume rule-based reasoning (programming) (Brown & O'Leary, 1995).

In the rule-based programming model, rules are used to represent heuristics, which indicate a set of actions to be performed in a given situation. A rule consists of an *if* portion and a *then* portion. The *if* portion of a rule is a series of patterns which specify the data (or situation) which cause the rule to be applicable. The *then* portion is the set of actions to be executed when the rule is valid (Riley, 2004). To construct an expert system capable of solving problems in a given domain, a knowledge engineer is needed. He/she starts by reading domain-related literature to become familiar with the issues and terminology. Then extensive interviews are held with one or more domain experts to obtain their knowledge. Finally, the knowledge engineer systematizes the results of these interviews and translates them into software (Schmuller, 1992). If an expert system uses fuzzy logic instead of Boolean logic, it is called a *fuzzy expert system*, a system which is a set of membership functions and rules used to reason about data. Unlike conventional expert systems, which are mainly symbolic reasoning engines, fuzzy expert systems lean toward numerical processing (Horstkotte, 2000).

Many expert systems have been developed in the CBM and fault diagnosis problem area. Condition monitoring of complex equipment involves a number of diagnostic procedures that utilize rules and judgments (Luxhoslashj & Williams, 1996).

2.1.1.1 Applications of Expert Systems in Condition-Based Maintenance

There are too many rule-based expert systems developed for fault diagnosis to consider all of them here, but a few key examples follow. Autar (1996) describes the development and implementation of an automated diagnostic expert system for diesel engines. This system is based on artificial intelligence criteria that use mechanical signature analyses (MSA) of signals acquired from engine mounted sensors. Based on the values of explanatory variables such as vibration signals, oil pressure and temperature, crankcase pressures, exhaust gas temperature and pressure, exhaust emissions, manifold noise levels, inlet manifold pressure, fuel delivery pressure, and instantaneous engine speed, the system can provide consistent diagnostic advice. Kawahara et al. (1998a, 1998b) propose a supporting expert system for planning of outage and retaining high power supply. The supporting expert system is based on the knowledge and experience of many planning engineers. In this case the system benefits from a number of related experts by aggregating their expertise resulting in more reasonable and consistent decisions. The proposed system has been applied successfully to test cases with 50 outage work requests taken from part of the 110 kV transmission line systems. Gale and Watton (1999) developed a real-time expert system for monitoring the condition of hydraulic control systems in a hot steel mill. They designed software architecture for building the predictive maintenance application based on a new signal-conditioning technique for detecting vibration in roll stack assembly. They also developed real-time feature extraction algorithms to classify the failure mode and effects related to the operational control of the mill. Gelgele and Wang (1998) propose a prototype to assist auto mechanics in fault diagnosis of engines by providing systematic and step-by-step analysis of failure symptoms and offering maintenance or service advice. The prototype has been developed using KnowledgePro, an expert system development tool that runs on a PC. and may well result in a systematic and intelligent method of engine diagnosis and maintenance. For more detailed review of the literature, see Badiru (1992), Capener et al. (1995), Holtzman (1988), Klein and Methlie (1990), Kumara et al. (1989).

Advantages: (Chee & Power, 1990; Luxhoslashj & Williams, 1996; Milne, 1996; Schmuller, 1992; Wegerich & Wilks, 2000)

- Unlike training a new human expert which is time-consuming and costly, making copies of an expert system is easy and cost efficient.
- Expert systems are consistent: Similar transactions are handled in the same way; comparable recommendations are made for identical situations.
- Expert systems can continuously monitor the condition of the system, unlike human experts who can not always be present to make expert decisions.
- Expert systems can report the rules and knowledge used to reach a particular solution and explain how such information was incorporated in the solution.
- Compared to neural networks, expert systems need less training and programming time.
- Expert systems can be built using fuzzy logic in cases where there are no strict *if* and *then* rules.

Disadvantages: (Autar, 1996; Chee & Power, 1990; Luxhoslashj & Williams, 1996; Milne, 1996; Schmuller, 1992; Wegerich & Wilks, 2000)

- Because expert systems are based on the expertise of human experts whose knowledge may be deficient, the expert system created based on that knowledge will carry the same deficiency.
- In expert systems, rules and knowledge can be changed relatively easily, therefore errors also can be introduced to the system more easily than for neural networks or statistical tools.
- Situations that human experts have not yet been exposed to cannot be dealt with in expert systems.
- Knowledge extraction is a very complicated process and only highly experienced knowledge engineers (facilitators as defined in Chapter 6) are able to extract and purify the required knowledge without bias.

2.1.2 Neural Networks

The neural network (NN) technique is inspired by the human brain's structure. Its learning capabilities have some similarities to those of the human brain (Hinton, 1992). The technique consists of looking for patterns in a set of examples and learning from those examples by adjusting the weights of the connections to produce output patterns. It can recognize patterns even when the data is noisy, ambiguous, distorted, or has a lot of variation (Brown & O'Leary, 1995). Neural networks consist of neurons that are connected. They have one input layer and one output layer between which are some hidden layers (Figure 1).

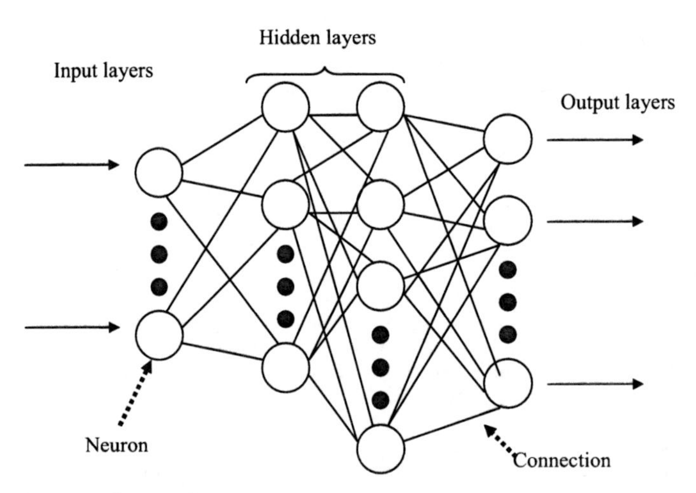

Figure 1: A neural network structure

The numbers of neurons, hidden layers and their organization determine the architecture of the NN. After choosing the architecture of a NN, a training process is necessary to develop a neural network that links the input pattern with the correct answer. A set of examples (training set) with known outputs (targets) is repeatedly fed into the network to "train" it. The process is continued until the difference between the input and output patterns for the training set reaches an acceptable value (acceptable error).

Several algorithms are used for training a network; the most common one is back-propagation. Back-propagation is completed in two phases:

1. First, inputs are sent forward through the network to produce an output.
2. Then the difference between the actual and desired outputs produces error signals that are sent "backwards" through the network to modify the weights of the inputs (Brown & O'Leary, 1995).

Figure 2 shows the structure of a neuron and the positions of the weights.

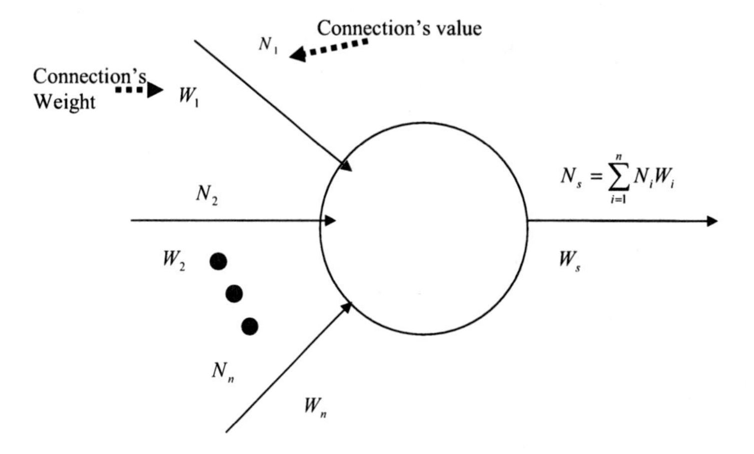

Figure 2: Structure of a neuron

2.1.2.1 Applications of Neural Networks in Condition-Based Maintenance

Neural networks have been used successfully in fault diagnosis in the areas of reliability and maintenance. For instance, to identify the state of a machine He et al. (1992) applied a multilayer feed-forward network. Knapp & Wang (1992) studied the application of a back-propagation network to fault-diagnose a CNC machine using vibration data. They studied network training efficiency by varying the learning rate and learning momentum of the activation function. Knapp et al. (2000) exploited a real-time based neural network to observe the condition of rotating mechanical equipment. They applied an ARTMAP neural network which continually screens machine vibration data as it becomes available to identify information about its condition. Their system was able to identify the presence of fault conditions with 100% accuracy on both lab and industrial data after minimal training; the accuracy of the fault classification was greater than 90%.

Recently, there has been considerable effort to combine fuzzy logic and neural networks for better performance in decision-making systems (Javadpour & Knapp, 2003). The uncertainties involved in the input description and output decision are taken care of by the concept of fuzzy sets while the neural net theory helps in generating the required decision region (Mitra & Pal, 1995). There is a large and growing body of literature (Buckley & Hayashi, 1993; Carpenter et al., 1992; Lin et al., 1995; Mitra & Pal, 1995; Uebele & Lan, 1995) pertaining to the fusion of neural nets and fuzzy systems.

Advantages: (Haykin, 1994; Hinton, 1992; Hopfield, 1982; Jardine, 2002; Ogaji & Singh, 2002; Patro & Kolarik, 1997; Sarle, 2002)

- NN appear to capture the nonlinearities in the data better than mathematical-function-based approaches.
- NN can tolerate non-repeatability problems or noise.
- Using condition monitoring data as input, trained NN can evaluate the system's health.
- NN are dynamic and can adapt to a new situation.
- In cases where hard and fast rules cannot be applied easily, NN can be combined with fuzzy logic (known as "fuzzy neural networks") to cope with the situation.

Disadvantages: (Haykin, 1994; Hinton, 1992; Hopfield, 1982; Jardine, 2002; Ogaji & Singh, 2002; Patro & Kolarik, 1997; Sarle, 2002)

- Usually the optimal network structure for a given problem is unknown.
- There are no well-defined criteria for the validation of a network.
- There is no guarantee for the convergence of training algorithms.
- No methodology can create information that is not contained in the training data. So NN cannot be used successfully in cases where the data were not previously introduced to the model.
- The structure of the NN must be chosen carefully, otherwise data can undo the training previously done by other data resulting in an un-training process.
- Because NN are adaptive and dynamic they adapt themselves to shifts in the data and eventually they assume the drifted condition as a normal condition
- NN have a drawback called *spillover*. This means that if a change happens to one variable its impact will be carried throughout the network making it seem that many variables have been changed even though only one actually has.
- NN normally do not include economic considerations and analysis when making condition-based decisions. This means they usually cannot answer the question: Is it more profitable to preventively replace or to run until failure?

2.2 Statistical Models in Condition-Based Maintenance

Most of the early techniques used in reliability and maintenance have been statistical tools. Statistical models purify, interpret and incorporate data based on certain procedures. In statistical models, data is transferred to some kind of knowledge based which appropriate decisions can be made. In the field of CBM, because it requires simultaneous measurement of different variables, more complex and elite models are needed. Among those listed by Gray et al. (1988) are: Probit analysis which deals with quantifying the relation between a stimulus and its response; log-linear models that measure the effect of explanatory variables on the lifetime of a system, and proportional hazards modeling (PHM) in which failure models based on explanatory variables are developed.

Probit analysis and log-linear models have only recently been introduced into the reliability and maintenance area, whereas proportional hazards models had previously been applied with significant success. Before considering more advanced statistical tools such as regression models, the simplest statistical tool in CBM, trending, will be introduced with some of its applications in industry. PHM will be addressed separately in Chapter 3.

2.2.1 Trending (SPC)

One of the first techniques applied in CBM is trending or statistical process control (SPC). In this technique a threshold value is defined for each explanatory variable. The health condition of the system is estimated by comparing the covariates' level with their assigned threshold values. These threshold values have different names and meanings such as warning limits, emergency limits, critical limits or alarm limits. Replacement or continuing operation decisions are based on the meanings of these limits. Usually a component should be replaced if one or more explanatory variables exceed their critical or emergency limit. See Figure 3.

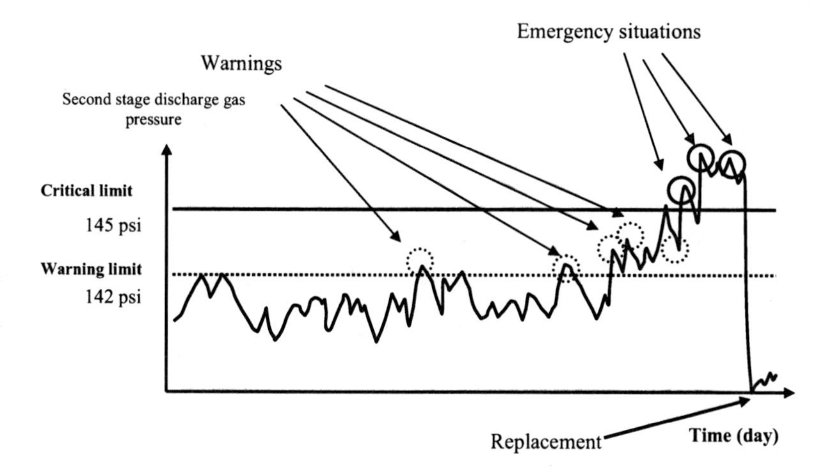

Figure 3: Statistical Process Control Chart for 2nd-stage discharge gas pressure of a reciprocating compressor

The SPC technique is easy to use (McCrea, 1992), is easily applied, and does not require much staff training. The greatest benefit can be obtained if the technique is applied online. Several steps are involved (Scarf, 1997): monitoring some condition-related variable(s) (Chen et al., 1994); designing condition monitoring data acquisition systems (Drake et al., 1995); diagnosing condition monitoring data (Harrison, 1995; Li & Li, 1995), and implementing computerized condition monitoring (Meher-Homji et al., 1994). For example, Crawford and Weinstock (1990) illustrate the NASA trend analysis program. Its four major parts (problem/reliability, performance, supportability, and programmatic trending) are described and examples from space shuttle applications are presented. The main focus is on the programmatic trending component of the program and several of the applied statistical techniques. Wach (2003) applied standard IEC 61502 (standard of the International Electro-technical Commission) as an example of early failure detection and on-line condition monitoring methods based on signature analysis and feature monitoring. IEC 61502 incorporated longstanding methods for a new type of nuclear power plant instrumentation and control system. This standard combined with powerful modern computer technology allows feature extraction and trending, alert level monitoring, and remote data signature transmission of dynamic process signals such as neutron noise, vibration signals, or pressure noise. There are many more applications and studies in this area, however it is outside of the scope of this research to elaborate further.

Advantages: (Crawford & Weinstock, 1990; Liptrot & Palarchio, 2000; Patro & Kolarik, 1997; Wach, 2003)

- It is very simple and easy to use.
- It uses only statistical analysis so no knowledge about the dynamics of the system is needed.
- It focuses on monitoring and observing all influential explanatory variables. This provides raw material for other techniques using condition monitoring data, such as PHM, NN, and expert systems.
- It can be applied online thereby replacing human labor, reducing costs, saving time, and resulting in consistent decision-making.

Disadvantages: (Crawford & Weinstock, 1990; Liptrot & Palarchio, 2000; Patro & Kolarik, 1997; Wach, 2003)

- SPC works based on the assumption that future observations are identically distributed and independent. Therefore, it cannot be applied successfully when data are correlated and dependent.
- SPC is a static tool. It consists of warning and emergency limits which do not vary with time and thus cannot take into account system "aging" which usually affects the meaning of the condition monitoring variables.
- SPCs are not capable of considering costing issues. They give the same response regardless of changes in the cost of failure or the cost of preventive maintenance.

2.2.2 Regression Models

Different regression models have been used to measure the effects of fixed explanatory variables on the health condition of a system. These models can explain or predict why some units/systems fail quickly and others survive a long time. The purpose of a regression model is to state the failure-time distribution as a function of **n** explanatory variables (covariates) denoted by: $x = (x_1, x_2, ... x_n)$. In this case:

$$P(T<t; x) = F(t; x)$$

Therefore the parameters of the system's failure-time distribution are functions of the explanatory variables.

$$\theta = (\theta_1, \theta_2, ... \theta_k) = (\theta_1(x_1, x_2, ... x_n), \theta_2(x_1, x_2, ... x_n), ... \theta_k(x_1, x_2, ... x_n))$$

Where θ is the model parameter vector.

An important class of regression models permits one or more of the elements of θ to be a function of $x = (x_1, x_2, ... x_n)$. In general, functions are used in a specified form with one or more unknown parameters that need to be drawn from data. In a general form of the statistical regression analysis the most commonly used models have the mean of the normal distribution depending linearly on a vector of explanatory variables ($x = (x_1, x_2, ... x_n)$) (Meeker & Escobar, 1998):

$$\mu = \beta . x^T = \sum_{i=1}^{n} \beta_i x_i$$

In some cases there may be more than one distribution function parameter that is dependent on one or more explanatory variables.

2.2.3 Accelerated Failure Time Model

The accelerated failure time (AFT) model is frequently used to express the effect of explanatory variables (***x***) on the failure time. In this model there is a simple time scaling acceleration factor that is a function of ***x*** (Meeker & Escobar, 1998):

$$T(x) = T(x_0) / F(x)$$

Where $T(x)$ is the failure time at condition ***x*** and $T(x_0)$ is the related failure time at some "baseline" conditions x_0. In many cases $F(x)$ is assumed to have the following log linear relationship:

$F(x) = 1/\exp(\beta_1 x_1 + ... + \beta_n x_n)$ Where $F(x) > 0$, and $x_0 = 0$.

When $F(x) > 1$, the model accelerates time in the logic that time progresses more quickly at x than at x_0 so that $T(x) < T(x_0)$. When $0 < F(x) < 1$, $T(x) > T(x_0)$, and time at x is slowed down relative to time at x_0. Different regression models can be built using different functional relations between $T(x)$ and x. In some models the correlation between different explanatory variables can also be considered. If the failure time is directly a linear function of the covariates, the model is called a *linear regression* model:

$$T(x) = \beta_0 + \beta_1 x_1 + ... + \beta_n x_n + \varepsilon \ ;$$

where ε is the error term.

2.2.4 Applications of Statistical Models in Condition-Based Maintenance

Shyur et al. (1999) applied a general regression model for accelerated life testing. The model uses failure time data at accelerated conditions to estimate the reliability measures at normal operating conditions. Not only has the model proved its applicability in the reliability engineering field, but it was also used successfully in analyzing the survival time data of non-homogeneous populations in the medical field.

Lee and Yang (2001) presented a new method using a multiple linear regression model for the fault diagnosis of temperature sensors along with the recovery of flawed data. Temperature sensors in a machine tool thermal error compensation system improved the machining accuracy by supplying reliable temperature data on the machine structure. The method's effectiveness was tested by comparing computer simulation results with measured data from a CNC machining center.

Chinnam (2002) developed a polynomial regression model for degradation signal modeling. The model was applied to monitor high-speed steel drill-bits (on a cutting tool machine) used for drilling holes in stainless-steel metal plates. In its second application the model was tested for modeling and forecasting fatigue-crack-growth data from the literature. The model was used to estimate and forecast the reliability of plates expected to fail due to fatigue-crack-growth. For both studies the results were seen to be very promising.

Krivtsov et al. (2002) developed an empirical approach to the root cause analysis of automobile tire failure. As the explanatory variables they used parameters related to tire geometry and physical properties that potentially affect a tire's life. Cox's regression model was used to analyze lifetime data. The paper also deals with the application of a linear regression to model the failure initiation and propagation.

Zhang et al. (2002) developed a model to measure the reliability of bearings. A bearing's reliability is normally established by repeated life testing which provides valuable information on the fatigue mechanisms from crack initiation, and crack propagation to flake. Accelerated life testing was applied to analyze the bearing's lifetime data accompanied by explanatory variables such as temperature, voltage, pressure, and corrosive media. The experimental outcomes illustrated that the accelerated life test model can effectively assess the life probability of a bearing based on accelerated environmental testing.

Advantages: (Chinnam, 2002; Cox, 1972; Jolliffe, 1986; Krivtsov et al., 2002; Lee & Yang, 2001; Meeker & Escobar, 1998; Shyur et al., 1999; Wegerich & Wilks, 2000; Zhang et al., 2002)

- To be operative, regression models need less data compared to NN and expert systems
- Regression models can define and calculate confidence intervals for their answers. This gives users a degree of certainty about the solutions provided by the models.
- Using Bayesian statistics, most of the statistical models can now incorporate experts' knowledge in the model, which gives them the advantage of using more sources of knowledge in decision-making.
- Regression models consider working age as a variable affecting the failure time of a system. In this sense they can provide more realistic solutions than SPC.

Disadvantages: (Chinnam, 2002; Cox, 1972; Jolliffe, 1986; Krivtsov et al., 2002; Lee & Yang, 2001; Meeker & Escobar, 1998; Shyur et al., 1999; Wegerich & Wilks, 2000; Zhang et al., 2002)

- Regression models are sometimes complicated and difficult for engineers to understand and work with.
- If the data is correlated, the model may provide narrow confidence intervals which can leave engineers and users overly confident about results that may not be reliable.
- Statistical models are static rather than dynamic. The explanatory variables are collected at discrete time intervals instead of continuously.

3. PROPORTIONAL HAZARDS MODEL

Introduced by Cox (1972), the proportional hazards model (PHM) was developed to estimate the effect of different covariates on the failure time of a system. The assumption in PHM is that the hazard rate of a system is the product of an arbitrary and usually unspecified baseline hazard rate $\lambda_o(t)$ dependent only on time and a positive function $\Psi(z;\gamma)$, mainly independent of time. The former incorporates the effect of time and the latter includes the effect of a number of covariates such as temperature, pressure, and changes in design (Kumar & Klefsjo, 1994).

Thus

$$\lambda(t;z) = \lambda_o(t)\Psi(z;\gamma)$$

Where z is a row vector consisting of (time-independent) covariates, and γ is a column vector consisting of regression parameters. The covariate z is associated with the system conditions and γ is the unknown parameters of the system. This was the original form of the PHM, however, through time some changes have been made to this format. For example different researchers use different functionality formats of $\Psi(z;\gamma)$. Some of these forms are: the exponential form, $\exp(z\gamma)$; the logistic form, $\log(1+\exp(z\gamma))$; the inverse linear form, $1/(1+z\gamma)$; and the linear form, $(1+z\gamma)$. Among these, the exponential format for $\Psi(z;\gamma)$ is most widely used because of its simplicity (Kumar & Klefsjo, 1994).

There are two assumptions regarding the baseline hazard rate:

1. Semiparametric PHM: This is the original form of PHM in which no assumptions are made about the nature or shape of the underlying baseline hazard. Under this assumption, the underlying baseline hazard rate is considered to be a step function, which has a jump just before the times to the failures (Kumar & Klefsjo, 1994).

2. Parametric PHM: In this form, a parametric function such as Weibull or extreme value is specified for the baseline hazard function. Although the parametric assumption imposes some restrictions on the model, it is usually more suitable for predicting the expected number of failures for a certain time interval (Jardine et al., 1987; Love & Guo, 1991).

There are two different assumptions about the covariates. Covariates are *time independent* if they do not vary during one history (the interval between installation/renewal and failure/suspension of the component or system of interest) and *time dependent* if they do vary.

In this book few assumptions are made regarding PHM. Time-dependent covariates are considered and a Weibull baseline hazard and exponential regression form as defined below are used:

$$\frac{\beta}{\eta}\left(\frac{t}{\eta}\right)^{\beta-1}\exp(\gamma_1 Z_1(t)+\ldots+\gamma_m Z_m(t)) \quad \textbf{(1)}$$

where:

$\frac{\beta}{\eta}\left(\frac{t}{\eta}\right)^{\beta-1}$ is a Weibull baseline hazard function with the parameters β and η

γ is the regression coefficient vector,

$Z(t)$ is the time-dependent covariates vector.

PHM has been applied in many fields of study including wide use in the biomedical field (Farewell, 1979; Gore, Pocock, & Kerr, 1984), finance (Henebry, 1996; Lane et al., 1986), demography (Diamond et al., 1986) and there has been an increasing interest in its application in reliability engineering (Gasmi et al., 2003; Jardine et al., 1987; Jardine & Tsang, 2006; Kobbacy et al., 1997; Kumar & Klefsjo, 1994; Kumar & Westberg, 1996; Kumar & Westberg, 1997).

3.1 Proportional Hazards Model in Reliability

In the paper written by Cox in 1972, he predicted that PHM would be used in reliability and medical science. PHM was first adopted by researchers in medicine to measure the effects of different conditions on the health and recovery rate of patients. The first application of PHM in reliability was in the 1980s when failure process of some components in a light water reactor in a nuclear plant was modeled using PHM by Booker et al. (1981). Ascher (1983) applied PHM to model marine gas turbine and ship sonar (see Kumar & Klefsjo,. 1994). Jardine et al., (1987) applied time-dependent PHM in the reliability area. They discussed the application of PHM in aircraft and marine engine failure data. With data from 27 engines, they used the Weibull PHM to measure the importance of factors such as flight hours since last overhaul and the levels of various metal particles in engine oil on engine overhaul time. Bendell et al. (1986) used PHM to model the time to failure of brake discs in high speed trains. Lindqvist et al. (1988) considered the failure mechanism of surface controlled subsurface safety valves (SCSSV) used in the North Sea. They collected the performance data of the SCSS valves from 26 oil and gas fields and used PHM to evaluate the impact of well parameters and valve characteristics on the reliability of the valves. Kumar et al. (1992) investigated the effects of two different designs and maintenance approaches on the reliability of a power transmission cable of an electric mine loader in an effort to illustrate the application of PHM in making decisions about selecting the proper material or the design of an item to meet the required purpose efficiently. Baxter et al. (1988) applied PHM to two subsets of a transmission failure and repair database. The influences of external variables such as weather, season, voltage, and time on the failure and repair data were investigated. For weather conditions, for example, two states were considered: normal and adverse. For season they considered a covariate that takes a value of 0 for summer (warm) and 1 for winter (cold). The model proved to be powerful and consistent. For a major weapon system, Gray et al. (1988) considered several analyzing methods such as: multivariate analysis, homogenizing the data and distribution fitting, time series analysis, and proportional hazards modeling in the analysis of early field data. PHM was chosen because of its considerable potential to handle this kind of problem.

3.2 Advantages of Proportional Hazards Model

PHM has almost the same advantages and disadvantages as other regression models (see section 2.2.4). However, what has made PHM more attractive is its ability to consider time-dependent covariates which most of other regression models cannot do. Unlike most other regression models, it is also capable of handling tied data and censored observations. (When the failure age is unknown, but it is known that an item has aged more than a specific time, that specific time is called a *censored observation*). PHM has proved to give more reliable results in real world applications (Gray et al., 1988).

In most early applications of PHM in the medical and biomedical sciences, all covariates were considered time independent. Thus, most of the early models using PHM did not deal with time-dependent covariates. However in the area of reliability there are many applications for time-dependent covariates, so some more recent studies have done so.

3.3 Likelihood in PHM with Time-Varying Covariates and Weibull Baseline Hazard

For a PHM with time-dependent covariates inspection records are obtained periodically to evaluate the state of the equipment. Data arrive in the following form:

$$(t_i, \delta_i, Z_1(t_i), Z_2(t_i), \ldots Z_m(t_i))$$

where t_i is the current observation time, $Z_1(t_i), Z_2(t_i), \ldots Z_m(t_i)$ are values of covariates at t_i, and δ_i is a dummy variable which denotes the current state of the history. If t_i is the failure time, δ_i=1; and if t_i is not a failure time (either inspection or suspension), δ_i=0.

The likelihood function of PHM with time-dependent covariates is complex because the values of covariates change during each history. Normally for each set of survival data there would be one likelihood component; however, PHM with time-dependent covariates requires one likelihood component for each inspection (see Table 1 and the example in section 4.2 to learn about the format of the data in a PHM with time-dependent covariates).

$$L(t_i, \delta_i, Z_1(t_i), Z_2(t_i), \ldots Z_m(t_i)) = \left(\frac{\beta}{\eta}\left(\frac{t_i}{\eta}\right)^{\beta-1} e^{\gamma_1 Z_1(t_i) + \ldots + \gamma_m Z_m(t_i)}\right)^{\delta_i} \exp\left(-\int_0^{t_i} \frac{\beta}{\eta}\left(\frac{s}{\eta}\right)^{\beta-1} e^{\gamma_1 Z_1(s) + \ldots + \gamma_m Z_m(s)} ds\right)$$

If t_i is an inspection time (temporary suspended time) when the next inspection happens at t_{i+1} the marginal contribution of this inspection record is:

$$\exp\left(- \int_{t_i}^{t_{i+1}} \frac{\beta}{\eta}\left(\frac{s}{\eta}\right)^{\beta-1} e^{\gamma_1 Z_1(t_i)+...+\gamma_m Z_m(t_i)} ds \right)$$

Here it is assumed that the value of the covariates for the inspection interval between t_i and t_{i+1} stays constant and is equal to $(Z_1(t_i),...,Z_m(t_i))$.

To arrive at the updated likelihood function, the previous likelihood function will be multiplied by the likelihood of the new data. The likelihood function after collecting *n* data $(((t_i, \delta_i, Z_1(t_i), Z_2(t_i),...Z_m(t_i))$, i=1, 2... n) will be (assuming data are end of history data) :

$$\prod_{i=1}^{n}\left(\frac{\beta}{\eta}\left(\frac{t_i}{\eta}\right)^{\beta-1} e^{\gamma_1 Z_1(t_i) + ... + \gamma_m Z_m(t_i)} \right)^{\delta_i} \exp\left(- \int_{0}^{t_i} \frac{\beta}{\eta}\left(\frac{s}{\eta}\right)^{\beta-1} e^{\gamma_1 Z_1(s)+...+\gamma_m Z_m(s)} ds \right)$$

To find maximum likelihood estimators (MLEs) of the parameters the Newton Raphson method can be used (Ben-Israel, 1966). This method gives answers in a relatively quickly compared to other methods. In this book this method is used to find the MLEs of the data where applicable.

4. BAYESIAN STATISTICS

According to Bernardo and Smith (2001), "Bayesian statistics offers a rationalist theory of personalistic beliefs in contexts of uncertainty, with the central aim of characterizing how an individual should act in order to avoid certain kinds of undesirable behavioral inconsistencies."

Below is a brief introduction to Bayes' theorem and some definitions that will be needed in this research.

4.1 Bayes' Rule

Let θ denote the unobservable vector of population parameters of interest (such as γ, α in formula (1)), and **y** denote the observed data (such as inspection or failure data, see also Table 1). Let $p(\theta)$ represent the distribution function (density function) of θ before seeing **y,** called *prior distribution.* "The prior distribution represents a population of possible parameter values from which θ of current interest has been drawn." (Gelman et al., 2004) In order to make probability statements about θ given y, a model providing a joint probability distribution for θ and **y** must be designed. The joint probability mass or density function can be written as a product of two density functions $p(\theta)$ and $p(y/\theta)$, referred to as *prior distribution* and *sampling distribution* (likelihood) respectively, as follows:

$$p(\theta, y) = p(\theta)p(y/\theta) \qquad \textbf{(2)}$$

Simply conditioning on the known value of the data **y**, using the basic property of conditional probability known as Bayes' rule, yields the **posterior density,** which is the distribution function of θ after observing **y**:

$$p(\theta / y) = \frac{p(\theta \cap y)}{p(y)} = \frac{p(\theta)p(y/\theta)}{p(y)} \qquad \textbf{(3)}$$

There are two kinds of prior distributions:

1. Informative prior distribution: This prior distribution carries some knowledge about the quantity of interest, e.g. knowledge about the location, standard deviation, and its behavior. According to Gelman et al. (2004) "The guiding principle is that we must express our knowledge (and uncertainty) about θ as if its value could be thought of as a random realization from the prior distribution."

2. Non-informative prior distribution: If a prior distribution is tried to represent lack of knowledge about the quantity of interest it may be called *non-informative prior distribution.* A complete non-informative prior distribution does not exist (Irony & Singpurwalla, 1997), however some such as uniform distribution and Jeffreys' prior (Jeffreys, 1961), provide an acceptable level of ambiguity about a parameter in most situations. In a non-informative prior distribution, the contribution of the data in constructing the posterior distribution should be dominant. Non-informative priors try to make the type of prior knowledge which, for a given model and for a particular inference problem within this model, would make data dominant (Irony & Singpurwalla, 1997). An example of a non-informative prior distribution is a uniform distribution although some researchers argue that it is not completely non-informative. Examples of both informative and non-informative prior distributions are presented in section 6.1.

4.2 Bayesian Posterior Distribution and PHM

The following example shows how a posterior distribution for PHM with time-dependent covariates is calculated in the conventional way:

Example:

Assume that the prior distribution functions for parameters in a case where we have only one covariate are as follows:

$$\ln(\beta) = Normal(1, 0.15)$$

$$\ln(\eta) = Normal(16819, 4515)$$

$$\gamma_1 = Normal(0.104951, 0.005316)$$

The joint prior distribution for the parameters assuming that they are all independent is:

$$P(\theta) = f(\ln(\beta), \ln(\eta), \gamma_1) = f_1(\ln(\beta)) \times f_2(\ln(\eta)) \times f_3(\gamma_1) =$$

$$\frac{e^{-\frac{1}{2}\left(\frac{(\ln(\beta)-1)^2}{0.0225}+\frac{(\ln(\eta)-16819)^2}{20385741.93}+\frac{(\gamma_1\text{-}0.1049)^2}{(2.82555\text{E-}05)}\right)}}{3.600026 \times \sqrt[2]{(2 \times \pi)}^3}$$

Now assume that the following data are the inspection records for one history which ends with failure.

Table 1

Inspection and Event Data for One History

Age (hours)	Iron(ppm)	Event
100	10	Inspection
180	26	Inspection
250	30	Inspection
312	37	Inspection
480	46	Inspection
566	51	Inspection
632	55	Inspection
698	72	Inspection
745	94	Failure

In the above table ppm stands for parts per million. 1 ppm is equivalent to one particle of a given substance for every 999,999 other particles.

The likelihood of this history will be: $P(y/\theta) = L(y/\beta,\eta,\gamma_1) = \frac{\beta}{\eta}\left(\frac{745}{\eta}\right)^{\beta-1}.e^{94\gamma_1}.$

$$\exp(\frac{1}{\eta^{\beta}}(100^{\beta}(1-e^{10\gamma_1})+180^{\beta}(e^{10\gamma_1}-e^{26\gamma_1})+$$

$$250^{\beta}(e^{26\gamma_1}-e^{30\gamma_1})+312^{\beta}(e^{30\gamma_1}-e^{37\gamma_1})+$$

$$480^{\beta}(e^{37\gamma_1}-e^{46\gamma_1})+566^{\beta}(e^{46\gamma_1}-e^{51\gamma_1})+$$

$$632^{\beta}(e^{51\gamma_1}-e^{55\gamma_1})+698^{\beta}(e^{55\gamma_1}-e^{72\gamma_1})+745^{\beta}.e^{72\gamma_1})$$

Therefore the posterior distribution will be calculated as follows:

$$p(\theta/y) \propto p(\theta)p(y/\theta) \rightarrow f(\ln(\beta),\ln(\eta),\gamma_1/y) \propto f(\ln(\beta),\ln(\eta),\gamma_1)\times L(y/\beta,\eta,\gamma_1)$$

$$f(\ln(\beta),\ln(\eta),\gamma_1/y) = K\times f(\ln(\beta),\ln(\eta),\gamma_1)\times L(y/\beta,\eta,\gamma_1)$$

$$K = \frac{1}{\iiint_{All\ \ln(\beta),\ln(\eta),\gamma_1} f(\ln(\beta),\ln(\eta),\gamma_1)L(y/\beta,\eta,\gamma_1)d\ln(\beta).d\ln(\eta).d\gamma_1} \quad \textbf{(4)}$$

To sample from such a complex distribution function, numerical methods and computer simulation must be used (Gelman et al., 2004). One method that is widely used and applied in the practice is the Markov chain Monte Carlo (MCMC) method (Brooks & Roberts, 1998).

4.3 Markov Chain Monte Carlo

Before 1990, calculating posterior distributions for the parameters in Bayesian models was difficult except for the simplest models. Computing the posteriors for models that were more complex than say, one-way ANOVA (analysis of variance) models required someone with comprehensive training in Bayesian methods. Even with this training, it was difficult to perform the required calculations. However, during the last decade, Markov Chain Monte Carlo (MCMC)–based sampling methods for evaluating high-dimensional posterior integrals have been developed. These include: Monte Carlo importance sampling (Hammersley & Handscomb, 1979 (see Ibrahim, et al., 2001)), Gibbs sampling (Gelfand & Smith, 1990), Metropolis-Hastings sampling (Hastings, 1970), and many other mixed algorithms (Ibrahim, et al., 2001). Using these algorithms one gets a sample of the joint posterior distribution by sampling from a Markov Chain. Then given this sample of the joint posterior distribution one can calculate desired values of the posterior distributions. Two of the oldest types of Markov Chain Monte Carlo algorithms are the Gibbs sampler and the Metropolis algorithm. Of these, the Gibbs sampler is regarded as the best known MCMC sampling algorithm in the Bayesian computational literature. According to Besag and Green (1993) the Gibbs sampler is founded on the ideas of Grenander (1983) (see Ibrahim, et al., 2001). However the formal term was introduced by his coworker (Geman & Geman, 1984). Here the Gibbs sampler is applied to the example described in section 4.2.

Let θ be the parameter vector for the Bayesian model. In this case: $\theta = (\beta, \eta, \gamma_1)$

The algorithm is described below:

Start with a set of initial values $\theta^0 = (\beta^0, \eta^0, \gamma_1{}^0)$

Now we use the joint posterior distribution of the parameters and take one sample of β conditioned on $\eta = \eta^0$ and $\gamma_1 = \gamma_1{}^0$ and call it β^1. Then we use the same joint posterior distribution of the parameters and take one sample of η conditioned on $\beta = \beta^1$ and $\gamma_1 = \gamma_1{}^0$. The procedure is continued as below:

Given the m-th update, get the (m+1)-th sample by:

$$\beta^{(m+1)} \sim [\beta \,|\, \eta - \eta^{(m)},\ \gamma_1 = \gamma_1^{(m)}]$$

$$\eta^{(m+1)} \sim [\eta \,|\, \beta = \beta^{(m+1)},\ \gamma_1 = \gamma_1^{(m)}]$$

$$\gamma_1{}^{(m+1)} \sim [\gamma_1 \,|\, \beta = \beta^{(m+1)},\ \eta = \eta^{(m+1)}]$$

By continuing this procedure we can get as many samples from the posterior distribution of the parameters as desired. Based on these samples the sample mean, median, or standard deviation can be calculated. To learn more about the MCMC and its characteristics and properties see Brooks et al. (1998).

4.4 Why We Use Bayesian Statistics in Decision-Making

Bayesian statistics can be applied in at least two ways. In decision-making parametric models for which the parameters must be estimated before they can be used usually are employed. When there is insufficient or no data expert knowledge is used to find the distribution function of the unknown parameters used in the model. In the case shown here Bayesian statistics are used to update the prior distributions of the parameters which were obtained using expert knowledge. This updated distribution function is called the *posterior distribution function.* Afterwards it can be used to find point estimators of the parameters and use these estimates in the model. If we are interested in finding the value of a decision function such as $C(x,\theta)$, we can approximate the value of the decision function by $C(x,\theta / \theta=\hat{\theta})=C(x,\hat{\theta})$, where $\hat{\theta}$ is a point estimate of θ using its posterior distribution function and x is a vector of variables which affect the decision.

Sometimes we have no data whatsoever but we can estimate the parameters by using the prior distribution function. Say for example we want to calculate the net profit of selling a product as a function of price and quantity when the accurate value of the quantity is unknown. However based on market analyses done by the marketing department it is known that the sale quantity is normally distributed with a mean of μ and standard deviation of σ. Total revenue function C is described as below:

$$C(x,\theta)= x\times\theta$$

Then, since this function has a linear relation with θ, the expected value of the revenue would be:

$$\mathrm{E}\,(\,C(x,\theta)\,)= C(x,E(\theta))=C(x,\mu)=x\times\mu$$

When C is a linear function of θ, the previous method works well and the results are consistent with the next method explained in this section. The previous expression can lead to inaccurate results in cases where C is not a linear function of θ. For example, assume the following relationship between C and θ:

$$C(x,\theta)=(\theta - x)^2, \ x \text{ is a constant}$$

Approximating the above function with $C(x,\hat{\theta})=(\hat{\theta}-x)^2$ can be very inaccurate especially if $\hat{\theta}$ is close to x.

So far we have described one way of using Bayesian statistics. Now we turn to a second method of using Bayesian statistics to handle the above problem. With more calculation, we are able to obtain more precise answers by using all available information about θ which is represented in its distribution function. This way of using Bayesian statistics is very popular in the field.

Following what was suggested above we have:

$$C(x) = \int_{\theta} C(x,\theta) f(\theta) d\theta \qquad \textbf{(5)}$$

where $f(\theta)$ is the density function of θ.

In our previous example it would become:

$$C(x) = \int_{\theta} C(x,\theta) f(\theta) d\theta = \int_{-\infty}^{+\infty} (\theta - x)^2 \frac{1}{\sigma\sqrt{2\pi}} e^{-(\theta-\mu)^2} d\theta = \sigma^2 + (\mu - x)^2$$

For more comprehensive details please see Percy (2002).

4.5 Bayesian Inference in Reliability

According to Siu and Kelly (1998), reliability assessment (RA) is intended to estimate the probability and consequences of accidents for a facility or process. But it often involves the analysis of low-probability events for which few data are available. Bayesian parameter estimation techniques are useful because, unlike classical methods, they are able to incorporate a wide variety of information types. These can include expert judgment as well as statistical data in the estimation process. In addition to their ability to deal with sparse data, Bayesian techniques are appropriate for use in reliability because they are derived from the structure of subjective probability. This structure, which articulates probability as a subjective (internal) measure of event likelihood, is a major part of current theories on decision-making under uncertainty (Keeney & Raiffa, 1993). However, because of some features that differ from classical statistics, such as its definition for confidence intervals, Bayesian techniques were not used in the area of reliability until the late 1970s (Apostolakis, 1978; Parry & Winter, 1981). Since then, because of their capabilities as mentioned above, Bayesian techniques have been used widely in reliability and maintenance (Siu & Kelly, 1998).

4.6 Proportional Hazards Model and the Bayesian Approach

Ferguson (1973) analyzed non-parametric problems using the Bayesian approach. In doing so he used the Dirichlet process which was the first random probability measure used in a Bayesian context. The Dirichlet process plays the role of a prior distribution and generates a distribution function F(.).

Later, the Dirichlet process was used by a few researchers such as Ghorai (1989) to model nonparametric PHM in a Bayesian context.

Kalbfleisch (1978) used a Gamma process to model the cumulative hazard function of a PHM with time-independent covariates. He divided the survival function into k disjoint intervals and placed a gamma prior on it to derive the posterior distribution. Then, he estimated the regression parameters and the underlying survival distribution. However, the main objective was to find the posterior distribution of the hazard as whole, not individual parameters.

Hjort (1990) also considered a nonparametric format for PHM. He too assumed the covariates are time independent. In his paper he criticized the model developed by Kalbfleisch (1978) and proposed a Beta process as a prior for the cumulative hazard function. His model was able to handle right-censored data.

Gray (1994) examined Bayesian methods for investigating the amount of institutional variation in a multi-center clinical trial. He used PHM with time-independent covariates and put normal priors on the regression coefficients of the covariates and the main effect of treatment. Posterior distributions were calculated using Gibbs sampling.

The only significant work using the assumption of time-dependent covariates in this field is by Gelfand and Mallick (1995). They considered the semiparametric PHM with time-independent covariates and non-informative prior distributions including Jeffrey's prior. They assumed that both the baseline hazard and the regression part are monotone functions and used a mixture of Beta distribution functions to model them. They then suggested that the same procedure can be applied for a case of time-dependent covariates. However they did not solve the problem because of the calculation burden.

Faraggi and Simon (1997) used multivariate normal distribution for the prior distribution (informative prior distribution) of the regression parameters of time-independent covariates. This setting enables incorporation of experts' knowledge. They did not make any assumption for the baseline hazard function and focused on regression parameters only. To make it simple, they assumed that there is no correlation between the regression parameters and used a partial likelihood method to update the prior distributions. Based on practical results they concluded that when the ratio of the number of data histories to the number of covariates is less than 15, the assumption of normality of the regression parameters has limited accuracy.

Laud, et al. (1998) dealt with non-parametric PHM with time-independent covariates. They focused on including right-censored data in their model. They also analyzed covariates based on failure times.

Kim and Ibrahim (2000) chose Weibull and extreme value distributions for the baseline hazard function. They assumed that covariates are time independent. In Weibull PHM they used a gamma prior distribution for the shape parameter and a uniform improper prior distribution for the regression coefficients which caused the model to be unable to accommodate expert knowledge regarding the effect of covariates. However, they used sample data to update the prior distribution and used Gibbs sampling to find the moments of the posterior distribution. The results were close to what was obtained using the maximum likelihood method. Their model also has the ability to work with right-censored data.

The model proposed by Kim and Lee (2003) is able to handle left-truncated and right-censored data and considers only time-independent covariates. They used a process neutral to the right as the prior of the baseline survival function. Then they placed a finite-dimensional prior on the regression coefficients. This non-parametric format created the exact form of the joint posterior distribution of the regression coefficients and the baseline cumulative hazard function.

Considering all the above research, to our knowledge no research has been done in this field with all three of the following assumptions: time-dependent covariates, parametric PHM and informative prior distribution.

Those interested can see Ibrahim et al. (2001) for an extensive review on the subject, and Merrick et al. (2003) for research regarding the role of Bayesian semiparametric analysis in the area of reliability and maintenance. However as mentioned earlier none of these works is directly applicable to the problem addressed in this book.

5. KNOWLEDGE ELICITATION TECHNIQUES

There are numerous techniques for eliciting knowledge in such fields as psychology, business management, education, counseling, and cognitive science (Cooke, 1994). These techniques are categorized in several ways such as quantitative versus qualitative techniques, or direct versus indirect (Oslon & Bioslsi, 1991). They are also categorized on the basis of their methodological similarities. Based on the latter, knowledge elicitation techniques are divided into the following three families:

1- Observations and interviews

2- Process tracing

3- Conceptual techniques

Techniques in the first family are usually more direct than the others. They have been used frequently by knowledge elicitors and provide a primary understanding of the domain problem. Techniques categorized under the second family are more specific and predetermined and are more suitable when deeper knowledge is needed. Techniques such as verbal reports and eye movements fit in this family. The last family includes techniques that produce representation of domain concepts and their structure or interrelationships. They are more indirect compared to other techniques. Conceptual techniques are not used in this research.

Before reviewing the concepts and literature on knowledge elicitation it is necessary to define some terminology:

Knowledge elicitation techniques are developed to elicit the knowledge of experts. The question is whom do we call an expert?

Garthwaite et al. (2005) define an expert as a person to whom society and/or his or her peers attribute special knowledge about the matters being elicited. In some circumstances we are seeking knowledge from people who are not regarded as experts according to the above definition. For example, when we ask people for their view of "gay marriage" we do not consider them experts; however, we need to ascertain their beliefs. Therefore, we need to redefine the concept of expert to consider an expert as a person whose knowledge we wish to elicit. The one who elicits the knowledge of an expert is called an *elicitor*, *facilitator* (as defined in Chapter 6), or *knowledge engineer* (as defined in section 2.1.1).

5.1 Observations and Interviews

The methods in this category are mostly informal and are suited for the initial phases of knowledge elicitation, especially since in most of them the elicitor does not need to have comprehensive knowledge about the domain. The downside of this technique is that the data thus gathered is usually difficult to interpret.

5.1.1 Observations

Observation is one of the most powerful elicitation techniques especially considering the unreliability of verbal techniques. Much can be learned by watching an expert perform a task or solve a problem (Welbank, 1990). Observations can be used to elicit knowledge that is hard to verbalize such as problem solving strategies or automatic procedures and it is useful to verify an expert's explanations while watching him/her work (Cooke, 1994).

The result of observations can be recorded in writing, audio or video. To know more about techniques and methods that can be used in recording and interpreting observation data, see Drury (1990). According to Hoffman (1987), familiar tasks, simulated familiar cases, limited-information tasks, constrained-processing tasks and tough cases can be targeted by observation. Familiar tasks are those experts are comfortable with and that can be performed in day-to-day work life; these tasks may represent the basics of work. In some cases however, task performance is costly or inflexible; in these cases, an expert can handle the task in a simulated environment and with archival data. In a limited-information task, the expert has little or no information about what is needed for the job. Sometimes the expert is limited by time or other factors; tasks performed under these conditions are considered to be constrained-processing tasks. Finally, tough cases are defined as those involving the most difficult tasks in terms of conditions, requirements and available information.

An elicitor can observe the expert in different ways. In *active participation*, the elicitor gets involved in the task while observing the expert. In *focused observation*, the expert focuses only on a small portion of the job. If the elicitor knows the aspects of the environment that need to be observed, *structured observation* can be used.

Although observation interferes least with the expert's task, his/her performance may be affected by the presence of the observer. Observation is also a slow method as it takes a long time to elicit all the required information. However, it is a valuable method for checking the validity of verbal reports. What the expert has explained verbally and what he/she meant by what he/she has said can be better understood by observing the expert in the work setting (Welbank, 1990).

5.1.2 Interviews

Interviews are the most frequently used method (Olson & Rueter, 1987). In general they are retrospective, which means that the interview takes place after the expert has performed his/her task. Two types of interviews exist: unstructured and structured.

5.1.2.1 Unstructured Interviews

Unstructured interviews are free-form. This means that neither the content nor the sequence of the questions has been defined before the interview. The elicitor asks very general questions and allows the expert to answer them in any way that he/she is comfortable with (Welbank, 1990). Unstructured interviews seem best suited for early knowledge elicitation sessions, for getting a broad view of the domain, but require some skills from the elicitor to facilitate the conversation. It is a stressful approach for the elicitor who has to listen carefully, to keep track of what has been said, and to keep the conversation going by asking useful and appropriate questions. Also, not everyone is comfortable with talking and expressing their opinions, and some people have an even harder time participating in a technical conversation without a format or structure (Welbank, 1990). Despite these criticisms about the efficiency of unstructured interview, a survey done by Cullen and Bryman (1988) found that 23% of participants used unstructured interviews and it was ranked as the most used technique.

5.1.2.2 Structured Interviews

Structured interviews differ from unstructured interviews in that they follow a predefined format (Schweickert et al., 1987). They can range from highly structured to semi-structured.

Since the knowledge elicitation process is more systematic in a structured interview, it has been viewed as a more complete and thorough method of gathering complete information in the domain than an unstructured interview. However, this organized and systematic method requires a lot more preparation and the elicitor needs to be more familiar with the domain knowledge in order to be able to design questions.

Several techniques can be used in a structured interview. If the focus is on a specific type of information such as cases, goals or diagrams, *focused discussion* can be used. Focused discussion is divided into three categories: case study analysis, goals and diagrams. In case study analysis, the focus is on specific experience and incidents (Bell & Hardiman, 1989). The expert can be asked to go through a simulated case similar to the real world condition. Therefore, the selection of cases is a critical condition in case study analysis.

Another method of conducting a structured interview is through *teachback* (Johnson & Johnson, 1987). In this technique the expert teaches the elicitor the knowledge that the elicitor is looking for and asks him/her to repeat it. This process continues until the expert feels the knowledge transfer is complete. In some modified forms of the teachback technique the expert gives some lectures about the domain knowledge at the beginning.

One advantage of this technique is that the expert himself gets trained when the elicitor presents the knowledge back to him. In cases where the expert has not organized his thoughts he might receive contradicting responses from the elicitor. So this method forces the expert to organize his thoughts and also encourages him to be more cautious when expressing his uncertainties about the concepts and variables.

Experts' knowledge can be elicited by asking questions and filling out questionnaires. The *questionnaire* is one of the widely used techniques for eliciting required knowledge. In this category of structured interview, open-ended questions regarding concepts and relations should be included (Cooke, 1994). Most of us have been asked to fill out questionnaires many times in our lives. According to the lecture notes (Milgram, 2003) for Course MIE 1403F (Methods in Human Factors Research, taught by Milgram): "a questionnaire is an instrument used to study a research problem, consisting usually of a collection of questions or statements given to a sample of individuals. Questionnaires are most often used in survey research." For more details on designing questionnaires see Olson & Rueter (1987) and Sinclair (1995).

Other methods of knowledge elicitation that can be categorized under the structured interview include: *role play* (Cordingley, 1989), *20 questions* (Breuker & Wielinga, 1987; Cordingley, 1989), *cloze experiment* (Cooke, 1994), *Likert scale items* (Likert, 1932), *question answering protocol* (Graesser & Clark, 1985), and *group interview techniques* (Cooke & Schvaneveldt, 1988).

These techniques can be used in different domains depending on the problem and situation. For instance, Boy (1997) has introduced a version of the group interview, the group elicitation method, to gather information for usability testing or product development. In this method, people from different groups who have an interest in the topic gather for a one-day session. The participants first write their ideas about the topic on a piece of paper. The papers are circulated and each person comments on others' ideas. This brainstorming continues until everyone is heard from and their viewpoint discussed. This method can be used in product development or consumer behavior analysis of marketing, or in other areas such as training and requirement-gathering (Boy, 1997).

5.1.3 Task Analysis

The last elicitation method in the observation and interview category is task analysis. This technique can provide insight about a specific task in a domain. Task analysis uses interview and observation, and breaks down tasks to analyze their behavioral implications (Wilson, 1989). There are different approaches to collecting data using task analysis techniques. Some, such as function flow analysis, operational sequence analysis, and information flow analysis, have been discussed by Meister (1989).

5.2 Process Tracing

Process tracing is more formal than the previous category and is concerned with both elicitation and analysis of knowledge. Its focus is on specific tasks and is aimed at collecting data about tasks during their performance (Elstein et al., 1978). The challenge here is to select tasks wisely so that only tasks that are representative of the overall process are chosen .

5.2.1 Verbal Reports

Verbal reports are the most common technique used in the process tracing family. In this technique the elicitor relies on knowledge that can be verbalized by the expert (Bainbridge, 1999; Cordingley, 1989). Verbal reports are usually used to collect data that helps to understand mental behavior (Bainbridge, 1999). Experts use a lot of mental activities in performing their tasks which cannot be observed but are a critical part of the knowledge they posses (Bainbridge, 1999).

There are some limitations listed by different scholars that should be considered before applying the verbal report method. First, with this method it would be hard to distinguish between explanations and the task if the task itself requires verbal communication (Bainbridge, 1999). Also, some tasks require the full attention of the expert making it impossible for the expert to talk while performing the task. Second, sometimes the expert does not explain what is obvious or routine (Bainbridge, 1999). One way to address this issue is to collect data by verbal report when two experts are talking to each other or in a mentoring program. Third, people differ in their ability to verbalize and they describe events differently. To minimize distortion in verbal reports the elicitation should be carried out in an environment that does not constrain what is said (Bainbridge, 1999).

Verbal reports can take the form of *verbal on-line reports* (Ericsson, 1984). In this method the expert provides the information concurrently with the task's performance. Some techniques that can be used in this method include self-reporting (Ericsson, 1984), which is a report of events by the expert him/herself, or shadowing in which another expert explains the actions of the one who is doing the task. In the self-report, the elicitor may ask the expert to think aloud, talk aloud or self-critique (Cooke, 1994).

In cases where the expert cannot comment while performing a task, a *verbal off-line* or *simulated recall* method can be used, in which the expert will explain the performed task after the task is accomplished (Elstein et al., 1978). Off-line reports can be written or verbalized (Bainbridge, 1999). For instance, the expert can fill out a questionnaire although this may limit the expert's answer. In questionnaires, questions have a set of predefined answers, either as percentages or rank of agreement, so the expert cannot express his/her opinion freely and it should fit it into these pre-defined categories. This problem can be solved if the elicitor is knowledgeable about the domain and is able to include all possible answers in the questionnaire (Bainbridge, 1999). To reduce reliance on the expert's memory, a video of the expert taken during task performance may be displayed to the expert.

5.2.2 Non-verbal Reports

Other non-verbal techniques can be used as well, especially when verbal communication is not possible. For instance, eye movement data (Carmody et al., 1984) can provide information about environmental cues. A record of keystrokes, facial expressions, and gestures are other sources of non-verbal data.

5.2.3 Protocol Analysis

The main purpose of data collection is to get an understanding of how experts make decisions. This information can be used in training programs or the development of expert systems (Bainbridge, 1999). Since analysis of data gathered by verbal and non-verbal reports is difficult, a protocol analysis technique can be used to help both data gathering and data analysis (Ericsson, 1984).

Content analysis (Bainbridge, 1999) organizes the collected database according to specific characteristics. For instance, the explicit protocol content can be analyzed by assigning each phrase to a category and the number of phrases in each category counted. Sometimes, the elicitor is interested in the implicit content of protocol which means some phrases contain information that are not verbalized explicitly. For instance, it might be inferred that the expert has some background knowledge in another field that affects his/her task performance (Bainbridge, 1999). *Discourse/conversation/interaction analysis* tries to identify different categories based on the interaction between the elicitor and the expert (Belkin et al., 1987).

5.2.4 Decision Analysis

Sometimes the elicitor is interested in the decisions made during a process and the reasoning behind those decisions; reasoning about such questions as: how many different variables and factors were considered and combined in the decisions? what are the consequences of the decisions? how likely is the occurrence of each incident? and what is gained by the decisions?

The decision analysis technique uses statistical methods and the theory of probability to quantify available information (Bradshaw & Boose, 1990; Slovic & Lichtenstein, 1971).

It should be borne in mind that there are cognitive biases in judgment, especially under uncertainty, that even experts cannot avoid (Hink & Woods, 1987). There are some de-biasing strategies that can be used to reduce the effects of an expert's bias (Cleaves, 1987; Mandi, 1988). As well, some have questioned the validity of studies based on poor expert judgment (Bloger & Wright, 1992; Keren, 1992; Shanteau, 1992).

5.2.4.1 Eliciting Estimations of Probability and Utility

A knowledge elicitor can gather required quantitative probability information about decisions made by an expert by asking the expert directly (Cordingley, 1989). The expert can list all the decisions, their consequences, the probability of occurrence of consequences and the profit gained by each consequence. The purpose here is to combine the decisions and calculate the worth of each decision with different statistical techniques such as Bayes' theorem (Slovic & Lichtenstein, 1971). The final goal is to select the decision that maximizes the worth. The result of this statistical analysis can be compared with the expert's selection. Sometimes the results differ, which shows that either the expert is using a different combination rule or that the probabilities have not been estimated accurately (Cooke & Schvaneveldt, 1988).

Note: More detailed explanation regarding modeling knowledge in a probabilistic format is given in Chapter 6. In this chapter, the literature review is based on the views of psychologists; whereas that described in Chapter 6 is based on the views of statisticians in the area of Bayesian statistics.

5.2.4.2 Statistical Modeling/Policy Capturing

In some areas, the expert needs to predict an event based on numerical data. In these cases, it is crucial to capture the decision-making pattern for further use and evaluation (Slovic & Lichtenstein, 1971). There are different statistical tools to capture how experts weight incidents. One tool is linear regression models (also explained in section 6.6.2.2) which build a relationship between relevant factors in the decision and the outcome of that decision. Such linear models can help predict future cases (Dawes, 1979). Another statistical method is analysis of variance (ANOVA) when variables are categorical rather than continual (Triggs, 1988).

5.3 Conceptual Techniques

This family of methods is focused on eliciting domain concepts and their interrelation and structure (Cooke, 1994). They are more formal and indirect techniques than those described above and require less verbalization. They are also good for cases where there are multiple experts and are capable of aggregating different experts' opinions to achieve one general structure. Since these techniques are only suitable for concept elicitation, it seems appropriate to combine them with techniques from the other two groups to ensure a full coverage of domain knowledge.

5.3.1. Data Collection Techniques

The elicitor is usually interested in the proximity and relatedness of different concepts and seeks to build a structure based on this information. This category of methods deals with building a matrix to show how different concepts of domain are related. Usually the results can be summarized across different experts and an aggregate matrix that represents the overall proximity of concepts can be built.

5.3.1.1 Rating and Ranking

A well-known method is rating and ranking. There are different techniques that can be used to achieve proximity of concept. For instance in *paired comparisons*, the expert gives a relatedness rank to every possible pair of concepts. Sometimes pairs of concepts are compared to a predefined pair and the expert is asked to rate the pairs in reference to the chosen pair; this is *magnitude estimation*. The downside of paired comparison is that it is time consuming, especially when the number of concepts is large. For instance, if the number of concepts is 30 or 40, 435 and 780 pairs of concepts can be formed respectively.

There are other methods, such as *controlled association*, in which for each concept all other concepts that have a relationship with it are selected (Miyamoto et al., 1986). Another, *reference ranking*, takes each concept as a reference point and ranks all other concepts based on their proximity and relatedness to that concept.

5.3.1.2 Repertory Grid

This is a more specific version of a rating technique (Cordingley, 1989; Shaw & Gianes, 1987) in which a set of dimensions is the focus of the rating. These specific constructs and dimensions may be predefined by the elicitor or extracted from the expert. By rating concepts along each construct a grid can be constructed.

5.3.1.3 Sorting

In sorting, the expert is asked to sort concepts into piles based on relatedness; the number of piles and number of concepts per pile is not important (Cordingley, 1989; Miller, 1969). Miller (1969) for instance asked several people to cluster a set of words. The number of times that two words were put in one pile by different people was taken as a measure of proximity and relatedness. Usually sorting is done several times with one expert or with different experts. This simple way of sorting is sometimes called *Q-sort.* If the sorting is carried out several times with the condition that at least one pile differ from the previous sort, then a *repeated sort* or *multiple Q-sort* is said to have occurred (Shadolt & Burton, 1990). In the *hierarchical sort procedure* the elicitor asks the expert to first sort concepts into two piles, then into three, and so on (Shadolt & Burton, 1990).

5.3.1.4 Event Co-occurrence/Transition Probabilities

In this technique indirect methods are used to achieve the proximity of data. The frequency with which two events co-occur, or the condition that one event happens given that the other has happened, is used as a measure of relatedness (Ekman & Scherer, 1982).

5.3.1.5. Correlations/Covariance

The relatedness and proximity of concepts can be represented by the statistical notation of correlation or measure of covariance. Different measures of each concept are needed to achieve correlation or a measure of covariance. These different measures may be obtained by asking different experts or asking the same expert about these concepts in different conditions (Cooke, 1994).

5.3.2 Concept Elicitation Techniques

It is essential to first define the important concepts (objects, elements, parts) of each domain. The methods described in this section are all intended to define these concepts. There are techniques that are concerned only with defining the terminology of the domain (Graesser & Murray, 1990), but others aim for more general definitions and the identification of concepts in a field.

5.3.2.1 Structured Interviews for Concept Elicitation

To elicit a list of concepts for a domain, techniques similar those discussed in the sections on structured interviews can be used. For instance, the expert can compare and contrast different concepts (their definitions, dimension, and application) and generalize the attributes of one group to another group.

Perhaps the easiest way to elicit concepts is to ask the expert to list the concepts, in a process called *concept listing*. If the expert is asked to list the steps involved in performing a task in the domain then the *step listing* technique has been used. Another way of extracting the concept of each domain is to ask the expert to assume he/she wants to write a book about the domain and ask him/her to name the chapters and sub-chapters of this hypothetical book. This is *chapter listing*, which is similar to the *lecture* method in which the expert is asked to present headings for a lecture (Gammack, 1987). Sometimes the concepts can be acquired by transcription of a conversation between the two experts regarding the domain. If this *interview transcription* is used, a third expert is needed to help in extracting concepts from the discussion transcription.

These methods differ in the number of concepts they elicit, their relevance, redundancy level, type (such as rules or fact), and the time required (Cooke, 1994).

5.3.2.2 Concept Elicitation Associated with Repertory Grid

As mentioned above in section 5.3.1.2, repertory grid is a method to analyze and organize the concepts of a domain. Different methods have been developed specific to eliciting concepts for building a repertory grid.

Laddering (Diederich et al., 1987; Shadolt & Burton, 1990) is one of these techniques which is similar to the structured interview and aims to extract super-ordinates, subordinates and attributes of each domain by asking question such as: why, what, how. *Triad comparison* or *triadic elicitation* (Cordingley, 1989) asks the expert to look at all triad combinations of concepts and select the one which is unlike the other two, and define the other two by a common attribute which is not in the third one. This method helps develop attributes of a domain which can be used in distinguishing different concepts.

5.3.3 Structural Analysis

The last technique in this family is structural analysis (Olson & Rueter, 1987; Shaw & Gianes, 1987). These are multivariate statistical techniques that aim to reduce pairwise relatedness estimations and offer a simpler method.

5.3.3.1 Multidimensional Scaling

This is a method which uses pairwise proximity of a set of concepts to generate a d-dimensional space (Shepard, 1962a; Shepard, 1962b). "Multidimensional scaling (MDS) is a set of data analysis techniques that display the structure of distance-like data as a geometrical picture" (Young, 1985). The objects or concepts are going to be represented by a point in the d-dimensional space and the distance between two points is defined by the amount of similarity that the two concepts have with each other (Young, 1985).

5.3.3.2 Discrete Techniques

These assume that concepts can be distinguished based on discrete attributes (Miller, 1969). Here the concepts are represented by clusters or graphics (Miller, 1969). Examples include the *clustering technique* which groups concepts based on the relatedness of members (Lewis, 1991), and *network scaling* in which concepts are represented by graphs based on their proximity (Cooke, 1994).

5.3.3.3 Direct Elicitation of Structure

Here the information elicited from a structured interview is represented in a graphical format such as a node/link network (Cooke, 1994). For instance, in the *graph construction task* method the expert is simply asked to explain the domain by drawing the concepts in a node/link relationship.

5.3.3.4 Interpretation of the Structure

This is used after the overall structure of the domain is arrived at. The interpretation may add more information, verify or modify the model and/or compare different structures with each other. These interpretations can happen through a *guided interview* in which the expert is asked to comment on the built structure.

5.4 Summary of Techniques

As mentioned earlier, different scholars have clustered knowledge elicitation techniques differently. Here the structure defined by Cooke (1994) has been selected as a main framework and the work of other scholars has been placed in this structure. The observation and interview family is the most informal category. As the methods move from informal to formal the elicitor becomes less active and less verbalization is needed. Using informal techniques requires the elicitor to have some training, for instance in interview skills, and may interfere with the expert's task; the data acquired in these ways are rich but need a lot of time to be analyzed. On the other hand, formal techniques need a different form of training such as protocol analysis skills and require more preparation on the part of the elicitor.

As shown, there are many elicitation techniques that can be used for different domains and different purposes. Selecting the appropriate elicitation technique(s) is a challenge, however, there are some basic guidelines for how to approach the elicitation process. For instance, the comfort of the expert should be considered (Welbank, 1990). Some experts are more comfortable with verbal methods, some prefer less verbalization. The expert cannot be expected to simply participate; some effort must be made by the elicitor to engage the expert and to gain his/her trust and interest (Welbank, 1990). The expert's comfort is of interest at the initial stage when trust has not been established and the elicitor and expert are not familiar and comfortable with each other (Welbank, 1990). Another factor to keep in mind is efficiency; different environments and domains need different techniques (Welbank, 1990). For instance in a surgery room, the verbal protocol is not going to work since the surgeon needs all the concentration and attention that he/she has for the surgery. Also as mentioned above, informal methods produce a lot of rich data but analysis of this data is hard; hence, it is more efficient to use these methods only in the first stages and to combine them with more focused methods.

However, the literature lacks an evaluation of these methods and an objective analysis of each method's usage; and most of the evaluations are based on intuitive or anecdotal evidence (Cooke, 1994). For instance Wagner et al. (2002) have evaluated knowledge elicitation techniques for the finance and accounting domain. Generally they have categorized domain problems into three categories: analysis problems (classification, diagnosis, interpretation, debugging), synthesis problems (configuration, design, planning, scheduling), and problems combining analysis and synthesis (command and control, instruction, monitoring, prediction, repair). Based on this classification of domain problems they have come up with some suggestions regarding the best knowledge elicitation technique. For analytic problem domains such as classification of asset write-down, a high degree of structure is desired, such as structured interview (Wagner et al., 2002). For synthetic and combination problems, the answer is not that simple. The elicitation method depends heavily on the complexity of the issue under investigation (Wagner et al., 2002).

It seems that the best way to overcome this lack of objective comparison is to use different techniques within each family and from different families to ensure coverage of all knowledge in the desired domain. Another way, in cases where it applies, is to make prototypes or test versions of the final product based on the information elicited from the expert and validate the prototype with the same expert or a group of experts. Although this method can also be used as an elicitation technique (Welbank, 1990), it seems more appropriate to use it for validation and verification. This process can be done when the prototype is finished or the expert can be involved in making the prototype from the outset as well as commenting on any missing or misunderstood information (Cooke, 1994).

6. TRANSFERRING KNOWLEDGE TO PRIOR DISTRIBUTIONS

Chapter 5 reviewed some of the many knowledge elicitations techniques that are applied in different fields of study. As discussed in sections 5.2.4.1 and 5.2.4.2, in some cases we are especially interested in quantification of that knowledge in the form of probability distributions. This is called *knowledge formalization* or *knowledge explication* as Cooke suggests (1994). In the area of Bayesian statistics it has several names: "elicitation" (Garthwaite et al., 2005; Kadane & Wolfson, 1998), "eliciting of probability distributions" (Garthwaite et al., 2005), "quantification of the expert opinion"(Garthwaite & Dickey, 1988), "elicitation of opinions" (Kadane & Wolfson, 1998), or "quantifying prior beliefs" (Chaloner & Rhame, 2001).

To be consistent with the majority of available literature, the term "elicitations" is used in this book, although it is a term used by psychologists to refer to the techniques described in Chapter 5. According to Garthwaite et al. (2005), "Elicitation is the process of formulating a person's knowledge and beliefs about one or more uncertain quantities into a (joint) probability distribution for those quantities."

In the context of Bayesian statistical analysis, elicitation is often considered as a methodology for specifying the prior distribution for one or more unknown parameters of a statistical model. This prior distribution will be combined with data using Bayes' rule to form the posterior distribution. This posterior distribution can be used in decision models. However, if there is no data, prior distributions can be used directly in decision-making. There exists a significant amount of literature on elicitation that deals with formulating a probability distribution for uncertain quantities when there are no data with which to enhance the knowledge articulated in that distribution. This situation occurs in decision-making where uncertainty about the "state of nature" needs to be expressed as a probability distribution to formulate and then maximize expected utility (Garthwaite et al., 2005). This is a common situation in the area of reliability since in many cases no data exist.

In some applications analysts use only point estimators of the parameters while in many others the full distributions of the parameters are used in calculating the decision functions. These situations were described in section 4.4.

In the literature of Bayesian statistics and formulating knowledge in probabilistic form, the definition of elicitor is different. Usually the term facilitator is used as defined below:

Facilitator (or elicitor as described in Chapter 5): Including a facilitator in the elicitation process is preferred. The facilitator's responsibility is to help the expert formulate his/her knowledge in probabilistic form. If the expert is a statistician and has enough background in statistics, there is no need for a facilitator. However this situation is very rare in practice. Elicitation is a very complex task requiring a range of skills and knowledge in order to be done properly. This fact makes the choice of a facilitator a crucial one.

Now the question is that when we can say that the elicitation is done well. According to Garthwaite (2005) an elicitation process is done well if the obtained distributions truthfully characterize the expert's knowledge despite the quality of knowledge. If the expert believes in a hypothesis strongly, the derived distribution has to reflect this by allocating a high probability for that hypothesis being true, even if it is subsequently found to be false. Even if most of the scientific community does not believe in the accuracy of that hypothesis, the distribution based on the expert opinions should support the accuracy of that hypothesis.

As mentioned earlier, performing an accurate elicitation is not easy or straightforward. The expert has to express his/her knowledge in probabilistic forms and therefore must be familiar with the basics of probability. Even for the knowledgeable expert it is difficult to assign probabilities to different events in many applications.

Assume that we are interested in eliciting a distribution for a continuous random variable X. To have a completely accurate elicitation we have to find an infinite collection of probabilities of $F(x) = P(X \leq x)$ for all possible values of x. This is not attainable in practice and usually experts can provide only a very limited number of such assessments. To show how knowledge can be transferred to probability distributions an example is presented in the next section. Please note that in this example the expert also plays the role of a facilitator. The example attempts to show how our knowledge changes based on new information and how it affects the probability distribution of the unknown variables.

6.1 Illustrative Example of How Knowledge Can Be Transferred to Prior Distributions

Try to predict the height of the next person who walks into your office. You will have a high degree of uncertainty about the figure's accuracy. But it is possible to guess the "typical" height of people who enter your office or to make a statement about the range of height that seems likely. Your belief may change with more information. For example, if told that the next person is a man, you would predict differently than if you were told it would be a woman. The same applies if we know that the next person coming to our office is a basketball player. The question is how can this uncertainty be shown mathematically. A common method suggests that we choose a known distribution function to describe the behavior of the variable of interest, which in this case is the height of the next person who walks into the office. Based on the rule of symmetry one can say that the distribution should be symmetric. Someone may even consider the normal distribution as being a good describer of the behavior of the variable which we will call X (height of the person). The normal distribution has two parameters, μ and σ. μ represents the location and σ describes the degree of scatter of the variable. μ and σ can be considered as elements of vector θ introduced in section 2.2.1. Now we have to assign values to μ and σ and also express our uncertainty about these parameters. For simplicity assume that σ is known and μ is the only unknown parameter. By assigning a value to μ and articulating our doubt, we in fact want to assign a distribution function to μ which is in contradiction with conventional statistics or non-Bayesian statistics. In non-Bayesian statistics we do not define any distribution functions for the parameters.

If we assume μ has a normal distribution function we have to estimate its mean and standard deviation which we call μ_0 and σ_0. These two new parameters are called *hyperparameters*. Based on our knowledge we may say that the average height of a person is about 165 centimeters. Therefore μ_0 is 165 cm. If someone asks for our certain opinion on the average height interval, we might answer [150-180]. This interval can be defined as $6\times\sigma_0$ therefore $\sigma_0 = \frac{180-150}{6} = 5cm$.

In summarizing the results so far we have:

$$X \sim N(\mu, \sigma)$$

$$\mu \sim N(\mu_0 = 165, \sigma_0 = 5)$$

The same logic works for σ although the mathematical burden may be greater.

We can say that μ has a normal prior distribution with a mean value of μ_0 and a standard deviation of σ_0. This is called an *informative prior distribution* since it carries some information about the parameter of interest, μ. If we reduce σ_0, the density function of μ becomes more clustered around its mean value of μ_0. This implies that we are more confident about the value of μ. Conversely, if we increase σ_0, the shape of the density function of μ becomes flatter. This means that our knowledge about the location of μ is less certain compared to the previous case. Asymptotically, if σ_0 goes to infinity it is equivalent to the fact that our knowledge about μ is approaching zero. Therefore we refer to a normal prior distribution with extremely high standard deviation as a non-informative distribution.

The above example considered parametric models which will be introduced in section 6.6.2. We will refer to this example later on when we review the literature on elicitation.

The elicitation described in the previous example would become very difficult if we wanted to include more than two parameters in the model. In that case the joint probability distribution of the parameters would have to be estimated, which is difficult. It would also be a time-consuming process if we wanted to estimate the value of the parameters with high precision. In that case we would have to do our best to remove any source of bias in the expert's opinion. For many reasons, there usually are some biases in a real elicitation. Some are general biases that are related to the way the human brain works. These are described in section 6.2. Some are based on other factors. For example, an expert might have some personal benefit in the results of the elicitation and might intentionally or unintentionally affect the results. As an example consider the estimated budget for different departments in a company for the coming year. Before the start of the next year each department is asked to give its best estimate of its expenses for the coming year. There is always a tendency for people in each department to overestimate expenses because a higher estimate might result in a larger budget allocation to that department. For more on this kind of bias, see Kadane & Winkler (1988).

Elicitation methods can be categorized differently. Some might categorize them as elicitation of parametric models versus elicitation of non-parametric models (Garthwaite et al., 2005). Some might divide them into *general* elicitation methods versus *application-specific* elicitation methods (Kadane & Wolfson, 1998). According to the latter, an elicitation method is general if it can be applied to a class of problems without any modifications. An elicitation method is application-specific if it can be applied only once; it is applicable only to the problem for which it is designed.

In spite of different categorizations all elicitation processes more or less follow the procedure suggested by Garthwaite et al. (2005):

1. *Setting up the elicitation*: this stage consists of preparing requirements of the elicitation including selecting the expert(s), training the expert, choosing an appropriate facilitator who has some knowledge regarding the domain of the problem, identifying what aspects of the problem to elicit, such as assessing possible sources of bias.
2. *Eliciting summaries*: in this stage the facilitator elicits specific summaries of the expert's distributions for the aspects defined in the set-up stage. This is the core of the elicitation process. To perform this stage well, one should have strong communication skills, be familiar with psychological issues, and posses advanced knowledge in statistics. That is why psychologists have contributed at least as much to this stage as statisticians.
3. *Fitting distributions*: in this stage the facilitator fits a (joint) probability distribution to the summaries. In practice this stage is usually closely connected to the pervious stage in that the choice of what

summaries to elicit is often influenced by the choice of what distributional form the facilitator intends to fit.

4. *Assessing the adequacy*: elicitation is always an iterative process, and at this stage, based on some measure of adequacy, the elicitation process has to be assessed. Depending on the results the process can return to the 2nd-stage.

Any knowledge elicitation protocol should more or less follow the above structure. However, in building elicitation protocols there are other factors that should be considered.

6.2 The Effect of Psychological Issues on Creating Bias and Eliciting Summaries

Elicitation transforms the expert's beliefs into some probability distributions. Therefore, it forms a bridge between an expert's beliefs and an expression of these beliefs in a statistically useful format (Garthwaite et al., 2005).

Hogarth (1975) stated "assessment techniques should be designed both to be compatible with man's abilities and to counteract his deficiencies."

Many authors have addressed psychological issues surrounding methodologies that quantify an expert's opinion. Much of the fundamental research was done in the early 1960s and 1970s by Hampton, et al. (1973), Hogarth (1975), Huber (1974), Lichtenstein et al. (1982), Slovic & Lichtenstein (1971), Tversky (1974), and Wallsten & Budescu (1983). There are also some more recent works in this area including the reviews done by Chaloner (1996), Cooke (1991), Hogarth (1987), Kadane & Wolfson (1998), Meyer & Booker (2001), Morgan & Henrion (1990), and Wallsten, et al. (1993). A summary of these psychological issues is presented in the next few sections.

6.3 Common Biases in Elicitation

There is always a question about how one evaluates the probability of an event or how one can conclude which of two or more events is the more likely to occur. It seems that the human brain uses some mental operations or heuristics to estimate the probability of events. Although these heuristics are very effective they can result in systematic bias and severe errors (Garthwaite, 2005; Meyer & Booker, 2001). One of these heuristics is judgment by representativeness. Possible biases based on this heuristic are discussed by Garthwaite et al. (2005), Kahneman & Tversky (1973), and Nisbett et al. (1976). This heuristic is related to questions of the form: "What is the probability that object A belongs to class B?"

In answering such questions, people usually compare the characteristics of A and B and according to their similarities, they assign a value to the probability of A belonging to B which in mathematical terms is P(B/A). The value assigned to this probability is usually proportional to the degree of the similarity between A and B, ignoring the unconditional probability of B (P(B)). A common example is called *librarian man*. Assume that Mr. X is described as meek, sober, meticulous, and reserved. Mr. X is working in one of the following jobs: salesman, pilot, physician, or librarian. If we ask a group of people to assess the probability that Mr. X is working in one of these positions, people overestimate the probability that Mr. X works as a librarian. This is due to the fact that they compare Mr. X's character with the stereotype of a librarian. However if we consider the fact that the number of librarians is much smaller than the number of salesmen, we see that the assigned probability to Mr. X being a librarian is higher than it should be.

Judgment by availability is another common heuristic. This heuristic deals with cases where a person is asked to estimate the frequency of a class or the probability of an event. Usually these probabilities and frequencies are assessed based on factors other than their true frequencies, such as the ease with which examples are recalled or occurrences come to mind. For instance, many people overestimate the probability of death as a result of an air crash, especially if such a crash has happened recently. Or a reliability technician might assign higher frequency to a catastrophic failure because such an event has happened and remains in his/her mind more strongly than cases where failures were detected in advance. This heuristic is discussed by Kahneman & Tversky (1973) in more detail.

Perhaps the most widely used heuristic when assessing probability of different events is judgment by anchoring and adjusting (Garthwaite et al., 2005). If a person is asked to assign a probability to an event and is a starting value for that probability, the starting value will affect that person's assessment. The starting value is usually called an *anchor*. An anchor could be suggested by the nature of the problem or the way that the problem is formulated (Garthwaite et al., 2005). In an experiment done by Tversky & Kahneman (1973) a group of people was asked to estimate the percentage of African countries in the United Nations. Some subjects were asked whether they thought this percentage is greater or less than 65%. Then they were asked to give their final estimation. The remaining subjects were given 10% as a starting value and were then asked whether they thought the true percentage is above 10% or below it. Those who were given 65% as their starting value ended up estimating the percentage of African nations to be 45% whereas the other subjects who had been given 10% as a starting value estimated African membership to be 25% on average.

There are other heuristics which are less common than those introduced here. For those who are interested, some of these are provided below, along with relevant sources:

Conservatism heuristic discussed by Peterson & Beach (1967),

Law of small numbers considered by Tversky & Kahneman (1971), and

Hindsight bias discussed by Fischhoff & Beyth (1975).

An extensive review of heuristics and biases in human judgment is given by Hogarth (1987).

6.4 Univariate Elicitation

In the elicitation process there usually is a choice as to the quantities to be elicited from the expert (Garthwaite et al., 2005). If possible, quantities should be chosen that are can be assessed more accurately and proficiently. The ability to assess simple statistical quantities such as mean, median, mode, standard deviation, and correlation has been tested in the field of psychology over several decades. These quantities are usually part of any elicitation process. Many researchers have investigated the ability of people to estimate sample proportions (Erlick, 1964; Nash, 1964; Pitz, 1965; Shuford, 1961; Simpson & Voss, 1961; Stevens & Galanter, 1957). In a series of experiments subjects were shown samples of binary data many times and were then asked to estimate the samples' proportions. In one of the experiments the subject were exposed to 20×20 square matrices. Some of the squares were painted blue, the remainder were red. In each trial a different proportion of red to blue squares was chosen. The subject was asked to observe the matrix for one second in some trials and for 10 seconds in others. At the end of each trial the subject was required to give his/her best estimate of the proportion of red squares to blue ones. Results showed that subjects were able to assess the sample proportion very accurately. In fact the difference between the mean of subjects' estimates and the true sample was less than 0.05 in most cases (Shuford, 1961).

Other experiments have measured the ability to estimate the central tendency of quantities' distributions (Beach & Swenson, 1966 (see Garthwaite et al., 2005); Peterson & Miller, 1964; Spencer, 1961; Spencer, 1963). In each trial a sample of numbers was presented to subjects. Later the subjects were asked to give their estimates for mean, median, and mode of the sample. When the sample distribution is symmetric these three numbers are very close to each other. In such cases the subjects were able to provide almost precise estimates for the three measures. In the case of non-symmetry they still could estimate median and mode with enough precision; however, their estimate for mean was biased toward median.

The results for other measures have not been as encouraging. In many cases we need to estimate the variances of unknown quantities to be able to construct their prior distributions. Unfortunately people's ability to estimate and interpret variances of unknown quantities is much less than their ability to estimate means and medians. It seems that they usually estimate coefficients of variations instead of variations of quantities (Garthwaite, 2005). For example, two sample data with the same variances and different mean values were displayed to subjects. The subjects provided a smaller estimate for the variance of the sample data with higher mean value (Lathrop, 1967 (see Garthwaite et al., 2005)).

A remedy for this problem is to avoid direct assessments of variances. Instead one can use credible intervals which are intervals that contain p% of the probability mass for the distribution of unknown quantities. These intervals combined with suitable distributional assumptions will enable us to estimate variances. Many experiments have tested people's ability in estimating credible intervals. If such experiments are executed well, the quantity of interest has to be inside a p% credible interval with a probability equal to p (Garthwaite, 2005). However, this is not true in many cases as there is a clear tendency for central credible intervals to be too small, so that probability is usually smaller than p. This bias is referred to as *overconfidence*, which means that people are usually more confident about the accuracy of their estimates than warranted. Some researchers have suggested that training the expert helps reduce this bias. To train experts, the elicitor must know the true values of the parameters and provide feedback to the experts if their estimations do not conform to the true values. However, in most practical situations elicitors do not know the true values. On the other hand, the results of such experiments show that although training experts will reduce *overconfidence* bias it would not eliminate it (Lichtenstein & Fischhoff, 1980; Schaefer & Borcherding, 1973). To learn more about this bias see Brenner, et al. (1996), Garthwaite & O'Hagan (2000), Murphy & Winkler (1974) and Peterson, et al. (1972).

Some researchers have designed innovative experiments such as visual tools to help experts express their knowledge more easily and with higher precision. For example, some researchers have used probability wheels which have shown some success in transferring uncertainties to prior distributions (Morgan & Henrion, 1990). A probability wheel consists of a round pie-shaped disc of one color that is partly covered by a "slice" of a different color and a pointer. The size of slice can be adjusted so that if the pointer is spun it lands within the slice with a probability equal to that of some specific events of interest to the elicitor (Garthwaite, 2005).

Some researchers have found interesting biases during elicitation experiments. *Effect of aggregation* is one. It occurs when an expert is asked to disaggregate an event to its constituent events (Johnson, et al., 1993). Fischhoff, et al., 1978, asked some car mechanics about the probable causes of a car not starting. They were asked to estimate the probability for a car not starting for reasons other than the battery, engine or fuel system. Their average estimated value for that probability was 0.22. Later they were asked to assess the probability that a car would not start for more specific reasons such as failure of the ignition system, failure of the starting system, etc, which are subcategories of the previous causes. Combining the probabilities of the latter disaggregated causes gave an average of 0.44 which is more than the probability they had given to the head category (0.22). Based on similar results obtained during different experiments there is now sufficient evidence to conclude that the sum of the probabilities for constituent events generally gives a much larger probability than a single assessment of the combined event that they form

(Garthwaite et al., 2005; Morgan & Henrion, 1990). For an extensive review on the topic see Tversky & Koehler (1994).

6.5 Multivariate Elicitation

Eliciting the expert's knowledge regarding more than one unknown variable is called *multivariate elicitation*. The output of such elicitation should be the expert's joint probability distribution for those variables. This type of elicitation is more complex than eliciting a distribution for only one variable. Types of questions, the structure of the elicitation, and the facilitator's job will be more complicated (Garthwaite, 2005).

A multivariate elicitation is much easier if it can be proved that unknown variables are independent. This means that obtaining new information regarding one variable does not affect the distribution of other variables. In this special case the multivariate elicitation will be broken down into several univariate elicitations. It also implies that the joint probability distribution of the variables is equal to the product of the marginal probability distribution for all variables. Fortunately, independence between variables can be judged moderately easily in many cases (Garthwaite, 2005). For cases where variables are not independent, researchers have developed several techniques to transform dependent variables into independent variables (Kadane & Schum, 1996; O'Hagan, 1998).

For example, assume that we are interested in eliciting the opinion of a medical expert regarding the effectiveness of two different treatments in a clinical trial. Let X and Y denote the relevant measures of effectiveness of the two treatments. For each sample of patients, X and Y are dependent in the sense that if X goes up the expert expects that Y goes up as well. The expert might relate the increase in the effectiveness of one treatment to variations in the quality of the patients in that sample and therefore might expect to see an increase in the value of the effectiveness of the other treatment as well. However, we can define $Z=Y/X$ as the relative effectiveness of treatment two over treatment one. If treatment one is a standard care and treatment two is a new treatment we can say that X and Z are independent (Garthwaite, 2005).

When there is dependency between variables the degree of dependency must be elicited. This usually is done by modeling dependency in terms of correlations. However, directly eliciting correlation coefficients or covariances may create many problems in complicated models. Clemen, et al. (2000) examined six different methods of eliciting correlation between weight and height in a population of male MBA students. Surprisingly they found that direct assessment of the correlation by specifying a value between -1 and 1 preformed the best among all six methods. These results are different from the results obtained in earlier experiments which concluded that direct assessment of the moments is a poor method of quantifying opinion (Gokhale & Press, 1982; Kadane & Wolfson, 1998; Morgan & Henrion, 1990).

Modeling multiple regression elicitations is another topic in the multivariate elicitation area that has attracted many researchers. It usually is concerned with finding the relationship between x-variables (generally called *cues*) and Y (usually called *criteria*). The aim is to estimate regression coefficients which are referred to as *cue weights*. Subjects predict the value of Y based on the known values of the cues. It has been found that in many cases the relationship between cues and criterion can be adequately described by a linear model. The correlations between subjects' responses and the responses predicted by linear models have usually been near 0.70 in real world experiments and in 0.80 and 0.90 in less complicated experiments or laboratory cases (Garthwaite, 2005).

Slovic & Lichtenstein (1971) provide an extensive review of cue-weighting experiments.

6.6 Fitting a Distribution

In this stage the facilitator converts the specific statements regarding unknown variable(s) obtained from the expert into a probability distribution. The fitting can be done at different levels of complexity. However, if the facilitator wants to use the Bayesian theorem to update the prior distributions to posterior distributions it is common to use parametric models (Garthwaite, 2005).

6.6.1 Uniform and Triangular Distributions

Experts usually can specify a range [a,b] which indicates where a parameter lies. If this is all that we know about that parameter then it appears sensible to use a uniform distribution over [a, b] to describe the distribution of that parameter. However, this view is too simplistic in the sense that the expert does not believe that the probability of observing the parameter at the edges (close to a or b) is equal to the probability of observing the parameter in more central parts (such as (a+b)/2). To remedy this problem some researchers suggested using a triangular distribution. For this purpose, they ask the expert to estimate a mode (say c) for the distribution of the unknown parameter. This results in a distribution of the following form:

$$f(x) = \begin{cases} 2\dfrac{x-a}{(b-a)(c-a)} & \text{if } a \le x \le c \\ 2\dfrac{b-x}{(b-a)(b-c)} & \text{if } c < x \le b \end{cases}$$

Some researchers have advocated using uniform and triangular distributions in some engineering applications (Oberkampf, et al., 2004) but O'Hagan and Oakley (2004) criticized this practice as failing to elicit experts' opinion adequately (Garthwaite, 2005).

6.6.2 Fitting Parametric Distributions

In some experiments the facilitator may find that an expert's knowledge can fit into a specified family of distributions. In such cases, the facilitator may decide to impose structure on the expert's knowledge by assuming that his or her opinion follows a parametric distribution. Each parametric distribution has one or more parameters (i.e., hyperparameters). The facilitator is responsible for choosing suitable values for hyperparameters based on the expert's knowledge. It is desirable to give higher priority to conjugate family distributions when choosing a parametric distribution for unknown quantities. This is very convenient when the unknown quantities need to be updated based on new data. In such cases the resulting distribution, *posterior distribution*, will be in the same family of distribution as the initial distribution (*prior distribution*, as defined in section 4.1). However as discussed in section 4.3, with new developments in computing one can use Markov Chain Monte Carlo methods and obtain samples from posterior distribution for many complex models even if the prior distribution does not belong to a conjugate family.

As a general principle, facilitators should ask experts about the quantities that are meaningful to them. This usually implies that the facilitators should seek observable quantities rather than unobservable ones. However in some cases it is difficult to find suitable observable quantities capable of providing required distributions. On the other hand in some applications, particular statistical models are so familiar to the experts that they can provide some information about their parameters directly (Garthwaite, 2005). Many such cases are discussed by Kadane (1980) and Winkler (1980).

Two elicitation tasks that have attracted significant attention are quantifying knowledge about a Bernoulli process and quantifying knowledge about a linear regression model.

The knowledge quantification procedure of a Bernoulli process is briefly explained below, followed by a discussion of linear regression models which are multi-parameter and have regression coefficients. These characteristics make them more similar to the proportional hazards model which is the main focus of this book.

6.6.2.1 Bernoulli Process

The first author to address problems in eliciting beliefs about a Bernoulli process was Winkler (1967). In a Bernoulli distribution we have only one parameter which is usually called p. In an example in his paper, Winkler (1967) assumed that p can represent the proportion of male students at the University of Chicago. In the literature four basic methods have been used to quantify subjective opinion about p. We describe one of them here, called the *quantile method.* To find out more about other models see Schaefer & Borcherding (1973) and Winkler (1972).

In the quantile method the facilitator asks the expert to specify his or her median estimate of p and also to provide one or more quantiles of his/her subjective distribution for p. Then the facilitator plots these points and plots a smooth cumulative distribution function through them. This is a non-parametric presentation of the expert's knowledge regarding p. If this non-parametric distribution can fit to a beta distribution the facilitator has to find the appropriate hyperparameters of the distribution. Beta distribution is the conjugate distribution for Bernoulli sampling and therefore it makes the updating process much easier. To facilitate the process of finding suitable hyperparameters, Winkler (1972) developed a table which lists several quantiles for a variety of parameter values.

6.6.2.2 Linear Regression Models

The normal linear regression models have been applied in many areas of research. Because of this wide applicability many authors have applied Bayesian statistics and knowledge elicitation techniques to normal regression models (Garthwaite & Dickey, 1988; Gill & Walker, 2005; Ibrahim & Laud, 1994; Kadane, et al., 1980; Oman, 1985). Here we illustrate the well-known basic multiple linear regression model as it has been described by Gill & Walker (2005) which has the following format:

$$\mathbf{y} = \mathbf{X}\beta + \varepsilon,$$

where **X** is a $n \times k$, rank k matrix of explanatory variables with a leading column of ones for the constant,

β is a $k \times 1$ vector of coefficients that has to be estimated,

$\mathbf{y}$ is an $n\times1$ vector of outcome variable values, and ε is an $n\times1$ vector of errors with the distribution function of $\mathrm{N}(0,\sigma^2\,\mathbf{I})$ for a constant σ^2.

The classical Bayesian approach to distribute priors for this model specifies: $p(\beta)\propto$ c over $(-\infty:\infty)$ for an arbitrary constant c, and $p(\sigma^2)=\frac{1}{\sigma}$ over $(0:\infty)$ (Tiao & Zellner, 1964).

So far we have considered noninformative prior distributions. This approach leads us to the marginal distributions for the parameters $(\mathbf{R},\beta,\sigma^2)$:

The covariance matrix, $\mathbf{R}=\dfrac{(n-k)\hat{\sigma}^2(\mathbf{X'X})^{-1}}{n-k-2}$

If $\mathbf{R}$ is a positive definite matrix, then: $(\beta\text{-}\mathbf{b})|\ \mathbf{X},\mathbf{y}\sim t_{n-k}$, and $\sigma^2|\mathbf{X},\mathbf{y}\sim$ IG(a, b).

IG(a, b) represents an inverse gamma distribution with a=(n-k-1)/2 and b=$\varepsilon'\varepsilon/2$.

However, in many situations we have experts with valuable knowledge regarding the parameters of the model making it worthwhile develop eliciting techniques that can enable that knowledge to be extracted. To quantify opinion about these quantities one could consider directly questioning the expert about the regression coefficients. Although some experts may be able to think directly of regression coefficients, as pointed out earlier, it is usually better to question people about observable quantities, such as $\mathbf{y}$, rather than asking direct questions about unobservable quantities, such as regression coefficients β (Garthwaite, 2005). Kadane et al. (1980) suggest the following approach:

Create m design points of the explanatory variable vector: $\widetilde{\mathbf{X}}_1,\widetilde{\mathbf{X}}_2,\ldots,\widetilde{\mathbf{X}}_m$, where these cases are interesting and common cases in the application. If we put all these variable vectors together, we create a $m\times k$ matrix which we call $\widetilde{\mathbf{X}}$. It is necessary that $\widetilde{\mathbf{X}}'\,\widetilde{\mathbf{X}}$ be a positive definite matrix.

Then, each assessor is asked to consider each of the $\widetilde{\mathbf{X}}_i$ row vectors and give his/her best estimate of $\mathbf{y}_{.50}$, a vector of medians of the outcome variables corresponding to the design cases. Therefore, based on experts' knowledge, point estimate for β is estimated by:

$\mathbf{b}_{.50}=(\widetilde{\mathbf{X}}'\widetilde{\mathbf{X}})^{-1}\widetilde{\mathbf{X}}'\mathbf{y}_{.50}$ which is quite intuitive.

To build the prior distribution for β we need more than just a point estimator. It can be shown that β has student-t distribution around $\mathbf{b}_{.50}$ with greater than two degrees of freedom. Having a degree of freedom greater than two ensures that the first two moments of the distribution exist (Kadane et al., 1980). This is considered a conservative prior distribution since large data size under weak regularity conditions leads to normal distributions for the coefficients after being updated to posterior distributions (Berger, 1993; Lindley & Smith, 1972). Unfortunately no one has provided any direct guidance as to how to estimate the degrees of freedom for this t-distribution since m value was established arbitrarily by the researchers. It also is not practical to elicit degrees of freedom directly from experts.

To solve this problem, Kadane et al. (1980) have suggested that for each $\widetilde{\mathbf{X}}_i$, experts should be asked to provide $\mathbf{y}_{.75}$. They might be asked to provide the median point above the median point ($\mathbf{y}_{.50}$) that was provided at the first stage. Following this procedure they can be asked for $y_{.9375}$. We then have to obtain the ratio defined below for all m assessments separately.

$$a(\tilde{X})=(y_{.9375}-y_{.50})/(y_{.75}-y_{.50})$$

The subtraction causes the numerator and denominator to be independent of the center. This ratio uniquely expresses tail behavior for some t-distributions. Kadane et al. (1980) have created a table which for different degrees of freedom shows their corresponding value of $a(\tilde{X})$. A subset of that table is shown below:

df	3	4	5	6	7	8	9	10	12
$a(\tilde{X})$	2.76	2.62	2.53	2.48	2.45	2.42	2.40	2.39	2.37

df	14	16	18	20	30	40	60	∞
$a(\tilde{X})$	2.36	2.35	2.34	2.33	2.31	2.31	2.30	2.27

Values greater than 2.76 indicate that the facilitator should instruct the experts to reevaluate their responses, and values less than 2.27 mean that a normal distribution can be used for β.

The case described here was one of the simplest cases of linear regression models. For more complex situations see the references suggested at the beginning of this section.

Other models that have been addressed include Weibull lifetime distributions (Singpurwalla & Song, 1987) and the proportional hazards model (Chaloner, et al.,1993).

6.6.2.3 Proportional Hazard Model

To the best of our knowledge, the model developed by Chaloner et al. (1993) is the only elicitation method that tries to estimate the regression coefficients of a PHM (Chaloner et al. in their paper explicitly say "There are no specific methods that we know of, however, for specifying a prior distribution on the regression coefficients of a proportional hazards model." In a private communication with Chaloner in June 2006, Chaloner said "As far as I know it is an open area ready for solution!") They considered a PHM with a time-independent covariate with a non-parametric assumption for the baseline of the hazard. Their method uses dynamic graphical displays of probability distributions that can be adjusted easily. This method is developed for a randomized trial comparing prophylaxes (prevention of disease or of a process that can lead to disease) for toxoplasmosis (an infection caused by a parasite that can lead to serious illness or death in the fetus) in a population of HIV-positive individuals. Prior distributions from five AIDS experts are elicited. Since the objective is to compare the effects of two different treatments (clindamycin and pyrimethamine) they use only two covariates. If the patient is receiving clindamycin treatment, the value of the first covariate is one and the value of the second covariate is zero, and vice versa if a patient is receiving pyrimethamine treatment. Each expert was asked to give his/her best estimates for the probability of experiencing the toxoplasmosis endpoint within the first two years of each treatment and for the case where a patient receives no treatment. Mathematically speaking, the expert was asked to provide his best guess for the following quantities:

$$P_0 = 1 - S(2)$$

$$P_p = 1 - S(2)^{\exp(\beta_p)}$$

$$P_c = 1 - S(2)^{\exp(\beta_c)}$$

In this formulation, the baseline hazard is assumed to be the same for each case. S(.) is the baseline survival function, and β_p and β_c are the regression coefficients for the two treatments (pyrimethamine and clindamycin).

To structure a parametric model, they have utilized a property of the proportional hazards model which indicates that the complementary log-log transformation of P_p and P_c has a range of the whole real line and is a linear function of the regression coefficients.

$$\log\{-\log(1-P_p)\} = \beta_p + \log\{-\log(1-P_0)\}$$

$$\log\{-\log(1-P_c)\} = \beta_c + \log\{-\log(1-P_0)\}$$

They have defined $\hat{P}_0$ as the best estimate for P_0 obtained using experts' knowledge, and have then used a parametric bivariate distribution for β_p and β_c conditional on $P_0 = \hat{P}_0$:

$$f(\beta_p, \beta_c \mid P_o = \hat{P}_o) = f(\beta_p, \beta_c)$$

Under the proportional hazards assumption the prior distribution should not depend on the interval; therefore, they have repeated their methodology by asking the expert about the cumulative probabilities of recovery after both two and three years and compared their results. Those results showed that the assumption of proportionality of hazard is not a suitable assumption in this case.

They assume that (β_p, β_c) have a type B bivariate extreme value distribution as given in Johnson and Kotz (1972) for $m \geq 1$:

$$f(x, y \mid m) = e^{m(x+y)}(e^{mx} + e^{my})^{-2+1/m} \times \{m - 1 + (e^{mx} + e^{my})^{1/m}\} \exp[-(e^{mx} + e^{my})^{1/m}]$$

Note that m=1 corresponds to independence of β_p and β_c. If μ_p and μ_c are location parameters, and scale parameters are σ_p and σ_c then the marginal distribution of P_p has the probability density function:

$$f(P_p \mid \mu_p, \sigma_p, P_0 = \hat{P}_0) = \frac{1}{(1-P_p)\{-\log(1-P_p)\}\sigma_p} \times \exp\{-\exp[\frac{\log\{-\log(1-P_p) - \mu_p}{\sigma_p}]\}$$

It shows that the corresponding distribution of $[-\log(1-P_p)]$ is Weibull and $\log[-\log(1-P_p)]$ is an extreme value.

To state the distribution on P_p and P_c the expert is asked to specify the upper and lower quartiles for P_p and P_c. These quartiles are used to find initial values for μ_p, μ_c, σ_p, and σ_c. The experiment starts with an assumption of independence (m=1). Then the expert is presented with plots of each marginal distribution and a dialog box with five sliders. The first four sliders can be used to adjust the four parameters μ_p, μ_c, σ_p, and σ_c. The sliders help the expert adjust the specified values interactively and observe the consequences directly through the marginal distributions for P_p and P_c. In their experiments they found that parameters β_p and β_c are dependent. In fact the fifth slider is used to adjust values of parameter m. Graphical feedback on the joint distribution is provided by a contour plot of the joint prior distribution of P_p and P_c. This contour plot has regions related to approximate by 20%, 40%, 60% and 80% density regions based on a χ^2 approximation to the log likelihood. Changing the value of m does not affect the marginal distributions but it changes the contour plot.

The results obtained from different experts differed substantially. For example four out of five experts estimated the value of $\hat{P}_0$ with 0.20, 0.75, 0.18, 0.65. Similar patterns repeated for regression coefficients, however, the differences were not as significant as they were for $\hat{P}_0$.

6.6.3 Non-parametric Fitting

Fitting parametric distributions to the expert's knowledge imposes some restrictions on the expert's underlying probability distribution. Although the distributional form may look plausible to the expert, the expert usually is not in a position to question the assumptions made by the facilitator. This problem is magnified in cases of multivariate distributions. As a result, there are many statisticians who are uncomfortable with parametric assumptions in modeling knowledge and data (Garthwaite, 2005).

A few non-parametric approaches are considered in elicitation. One approach is that the facilitator does not use any parametric distribution at all and just uses the expert's stated summaries. Bayes' linear method which is advocated by Goldstein (1999) (see Garthwaite et al., 2005) is one such method. This method is based on eliciting only first and second order moments which are means, variances, and covariances. Considering the fact that experts have difficulty assessing the moments (especially second order moments), the Bayes' linear approach requires high statistical understanding on the part of the expert. This means that the expert has to undergo substantial training sessions.

Another method is the one developed by Berger and O'Hagan (1988) (see Garthwaite et al., 2005). Based on their methodology the expert's prior distribution for a single unknown parameter is allowed to be any unimodal distribution having specified quantiles. For any given data they calculated the range of posterior inferences over this range of prior distributions. This method is considered to be fully non-parametric, however it is difficult to implement in more complex situations.

The last non-parametric fitting method discussed here is a recent approach developed by Oakley and O'Hagan (2002) which is in the framework of modern Bayesian non-parametric. In this method, the expert's beliefs about the random variable X are represented by a probability density function $f(x)$. From the facilitator's point of view, f is an unknown function and the facilitator has a prior distribution for f. This prior distribution is updated to a posterior distribution after observing some data or obtaining the expert's opinion. For the sake of simplicity we will not go into more detail.

7. PROBLEM DEFINITION

The proportional hazards model (PHM) with Weibull baseline hazard function (parametric distribution function) and time-dependent covariates is considered to depict the hazard rate of a system **(6)**. The objective is to estimate the parameters of this model ($\beta, \eta,$ and γ) using both expert knowledge and data by applying knowledge elicitation techniques and Bayesian statistics.

$$h(t;Z(t)) = \frac{\beta}{\eta}\left(\frac{t}{\eta}\right)^{\beta-1} e^{\gamma_1 Z_1(t)+\ldots+\gamma_m Z_m(t)} \qquad \textbf{(6)}$$

In the next few sections, four steps in this research are introduced along with their results. These results are obtained and/or tested and modified with a few experiments and a real industrial case. In solving the above problem the following points should be taken into account:

- The main objective is to develop a methodology that can be applied in industrial environments where most experts do not have a strong background in statistics. The methodology must be readily understood by those in the industry so that it will be appreciated and applied
- Descriptive functions of the parameters estimated based on an expert's knowledge should behave consistently with the expectation of that expert. For example, the PHM whose parameters are estimated based on the knowledge of an expert should provide mean time to failure, probability of failure based on different values of the covariates, ranking of different situations in terms of risk of failure, and other matters consistent with what the expert would predict based on his/her knowledge and experience.
- To be practical the knowledge elicitation process should not be time consuming since experts usually will not accept lengthy and extensive knowledge elicitation sessions.
- The prior distributions should be truly representative of the expert's knowledge. Making parametric assumptions for the prior distributions of the parameters of complex models such as a PHM imposes significant restriction on the parameters and as a result these prior distributions cannot embody the accurate knowledge of the experts.
- The process of obtaining prior and posterior distribution should be done in a relatively short period of time based on current computational facilities. Sometimes models are so complex that taking enough samples from their posterior distribution requires super computers or months of time.
- In cases where a number of people have some expertise regarding failure mechanisms of the same system, the methodology should combine expert knowledge from different sources. However, in the area of reliability such cases happen much less frequently than other areas using PHM, such as medical science.
- The results of knowledge elicitation and the updating process have to be considered within the context of reliability and maintenance in industry. To measure the reliability of equipment in industry, PHM is used to see when the hazard rate reaches a certain level. Replacement decisions are made based on some thresholds as explained in 2.2.1. Therefore our objective is to estimate the optimum time to replacement as accurately as possible. In this context paying extra attention to other aspects of elicitation that do not contribute to this objective would divert us from our main goal.

8. EXPERT KNOWLEDGE GATHERING

To estimate the reliability of a component as precisely as possible, one must use all related information from different sources. In industry experts' knowledge about the effects of time and covariate values $(t, Z(t))$ on the lifetime of a component is valuable. Some experts spend several hours a day working with machines or fixing them, others actually design the equipment. Thus they have considerable knowledge regarding deterioration of the components of the machine they are working with. Based on experience, experts can often determine the key factors that affect a failure process. They generally can compare the effects of those key factors with each other. They also usually have some information about average values of some variables, such as time to failure and condition indicators, and may be able to provide some confidence intervals for their estimations of those variables.

In order to build a practical model, its underlying assumptions should be driven by industries. To ensure this, there must be a continuous communication between theoreticians and practitioners during model building and subsequently at the model assumptions and model checking stages.

To fulfill this task, a list of questions that can be used to estimate the parameters of PHM was developed. To design questions that make sense to the people in an industry, several books containing industrial terminology were consulted (Campbell, 1995; Campbell & Jardine, 2001; Coetzee, 2004; Kelly, 2003; Moubray, 1997; Nyman & Levitt, 2001; Westerkamp, 1997; Williams, et al., 1994; Wireman, 1990; Wireman, 1994; Wireman, 2003; Wireman, 2005). Also broad research was done on the different tools and techniques applied in the area of condition-based maintenance. Techniques that use statistical models and those that use expert knowledge were studied and compared. Potential questions were shown to experts who have some industrial background to see if they make sense to people working in industry. A search was also carried out among software packages that use industrial expert knowledge to enhance the clarity of the questions for those working in an industrial environment. During this research five meetings were held with maintenance staff at Dofasco (a steel-producing company). Finally, more than 150 questions were put to these Dofasco workers via emails to test the results and knowledge elicitation process. Based on this work, the knowledge elicitation for the parameters of PHM can be broken down into the following stages:

1. *Unstructured interview*: in this stage the required information (explained in 8.1), such as the definition of failure, significant covariates, and mean time to failure, were collected in order to design more detailed questions (*Type 1* and *Type 2* questions as introduced in 8.3, and questions that measure upper and lower limits of parameter A, also introduced in 8.3).
2. *Teachback stage*: in this stage the information obtained in the previous stage was verified for consistency and accuracy (see 8.2).
3. *Customized interview and questionnaire*: based on the information obtained in the two previous stages and following the instructions presented in section 8.3.1, we developed *Type 1* and *Type 2* questions and questions that measure upper and lower limits of parameter A.

8.1 Unstructured Interview

At the beginning of the first meeting the objectives of the interview should be explained to the subject (expert), and the expert is invited to ask any question he/she might have regarding the process. Following this stage the expert will be asked some general questions regarding the process. This stage helps refresh the expert's memory and prepare him/her for more technical questions. Refresher questions could include:

- Please describe the maintenance activities you have regarding machine Y (in this case machine Y contains component X for which we are interested in building a PH model).
- How successful are your condition-based maintenance systems for machine Y?
- (Other)

Once the expert becomes interested in the interview more detailed explanations are sought and the questions zero in on component X. Sample questions could include:

- How do you define failure for component X?
- Which factors do you think are related to health condition of component X?
- Which of the above factors do you think is more important and why?
- What is the average time to failure for component X?
- Do you have any warning or emergency limits for these factors?
- If yes, how do you define them and what do they mean?
- (Other)

8.2 Teachback Stage

The information obtained during the interview is organized and presented to the expert to see if the expert agrees with it. This helps the expert spot possible mistakes and correct him/herself and/or the interviewer, and also gives him/her a stronger sense of collaboration and involvement in the knowledge elicitation process.

8.3 Customized Interview and Questionnaire

When the expert accepts the credibility and accuracy of the information presented to him/her, based on the acquired knowledge, more customized and detailed questions are designed. These are aimed at eliciting nuances of the process and must be designed and prepared carefully as they frame the knowledge elicitation methodology. The methodology used here is based on case comparisons and analyses. Each case describes a combination of a system's working age and covariate values. We define two cases in each question, Case A and Case B. There are two types of such questions. The first is called *Type 1*, the second *Type 2*. In *Type 1* questions the expert is asked to tell us in which case the system is at higher risk of failure, and to provide a confidence level for his/her estimation. In *Type 2* questions, we design the cases so that the second case (Case B) has a higher risk of failure. This can be done by making sure that all the factors (covariates and age of the system) that have an increasing effect on the hazard rate (this information is obtained during the earlier knowledge elicitation stages) either stay constant or increase from Case A to Case B; while those factors that have a decreasing effect on the hazard either decrease or stay constant from Case A to Case B (the details needed to design *Type 1* and *Type 2* questions are given in section 8.3.1). The expert is then asked to tell us how many times Case B is at higher risk of failure compared to Case A and to provide a confidence interval (usually 90% confidence interval) for his/her estimate. *Type 1* and *Type 2* questions are shown in the next two examples respectively. (Some real sample questions are brought in at the end of the case study which will be discussed in section 11.3.)

Example1 (*Type 1 question*)

Compare the following cases in terms of risk of failure:

(a) Use the sign “>“ if you think the component under the situation of Case A is at higher risk of failure.

(b) Use the sign “<“ if you think the component under the situation of Case B is at higher risk of failure.

(c) Use the sign "=" if you think that both situations have the same likelihood of failure for the component.

(d) Express your confidence about your answer on a scale of 0-100%.

Case A	Sign	Case B
Iron= 100 ppm,		Iron= 80 ppm
Chrome =70 ppm,	Confidence	Chrome =85 ppm
Age=200days		Age=200 days

Example 2 (Type 2 question)

How many times is case B at higher risk of failure compared to case A? Provide an interval that makes you confident about the answer.

Case A	Interval	Case B
Iron= 100 ppm,		Iron= 110 ppm
Chrome =60 ppm,	Confidence	Chrome =70ppm
Age=150 days		Age=200 days

Iron and chrome are covariates $Z_1(t)$ and $Z_2(t)$ in the PHM formula.

The answers to such questions (*Type 1* and *Type 2*) can be mathematically expressed using the hazard formula as:

$$b_l \le \frac{h(t_B, Z_B(t_B))}{h(t_A, Z_A(t_B))} \le b_u,$$

or

$$\ln b_l \le (\beta - 1)\ln(\frac{t_B}{t_A}) + \gamma'(Z_B(t_B) - Z_A(t_A)) \le \ln b_u, \qquad \textbf{(7)}$$

where $Z_A(t_A), Z_B(t_B)$ are vectors of the covariates corresponding to Case A and Case B respectively.

If in Example 1 (*Type 1* question) the expert says A has more risk of failure than B, this is equivalent to:

$$h(t = 200, z_1 = 100, z_2 = 70) > h(t = 200, z_1 = 80, z_2 = 85)$$

$$\Rightarrow \frac{\beta}{\eta}\left(\frac{200}{\eta}\right)^{\beta-1} e^{\gamma_1 100+\gamma_2 70} > \frac{\beta}{\eta}\left(\frac{200}{\eta}\right)^{\beta-1} e^{\gamma_1 80+\gamma_2 85}$$

$$\Rightarrow e^{\gamma_1 100+\gamma_2 70} > e^{\gamma_1 80+\gamma_2 85}$$

$$\Rightarrow \gamma_1 20 - \gamma_2 15 > 0$$

$$\Rightarrow \gamma_1 > 0.75\gamma_2$$

And if the expert in Example 2 (*Type 2* question) says case B is between two and four times more risky than A, this is equivalent to:

$$2 \le \frac{h(t = 200, z_1 = 120, z_2 = 105)}{h(t = 200, z_1 = 100, z_2 = 85)} \le 4$$

$$\Rightarrow 2 \le \frac{e^{\gamma_1 120+\gamma_2 105}}{e^{\gamma_1 100+\gamma_2 85}} \le 4$$

$$\Rightarrow \ln 2 \le \gamma_1 20 + \gamma_2 20 \le \ln 4$$

$$\Rightarrow 0.0346 \le \gamma_1 + \gamma_2 \le 0.0693$$

Using this method we can obtain upper and lower limits for β and all other combinations of $\gamma_i, 1 \le i \le m$. For example, $t_A = t_B, Z_{Bi} = Z_{Ai}$ for all values of i except for i=k will create an inequality only for γ_k.

The described methodology provides us with bounding inequalities of the following form:

$$a_0\beta + a_1\gamma_1 + ... + a_m\gamma_m \le c, \quad \textbf{(8)}$$

a_i, i=1,2,…,m, and c are real numbers, which defines the feasible parameter space.

To see how these inequalities define the feasible parameter space in simple case of having only two parameters, see Figures 4–6.

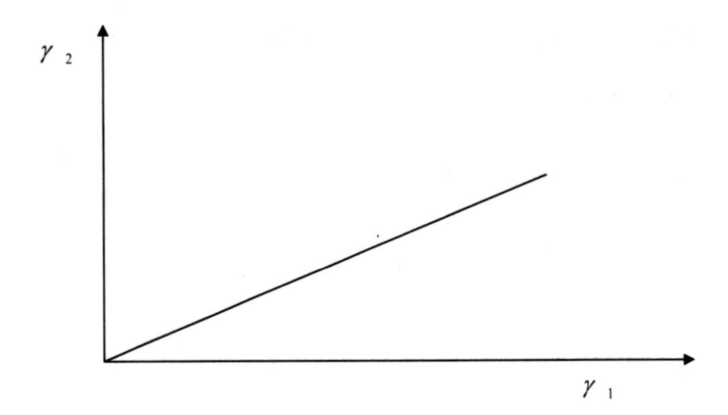

Figure 4: A line created by one Type 1 case comparison question

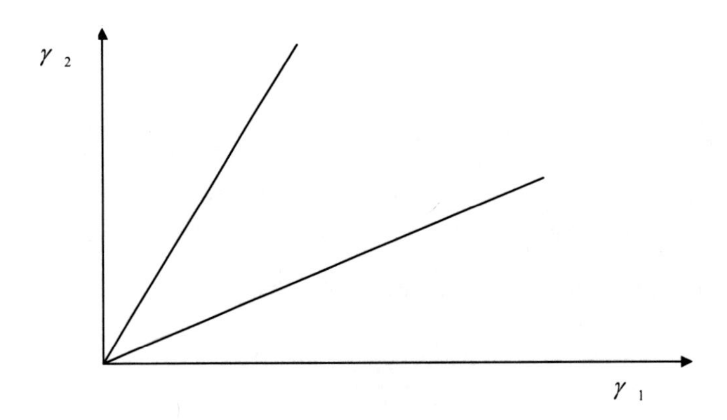

Figure 5: Two lines created as a result of two Type 1 case comparison questions

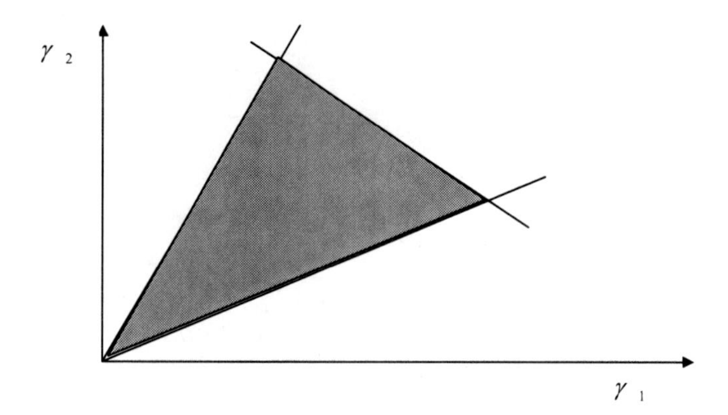

Figure 6: One Type 2 question and two Type 1 questions create a bounded area

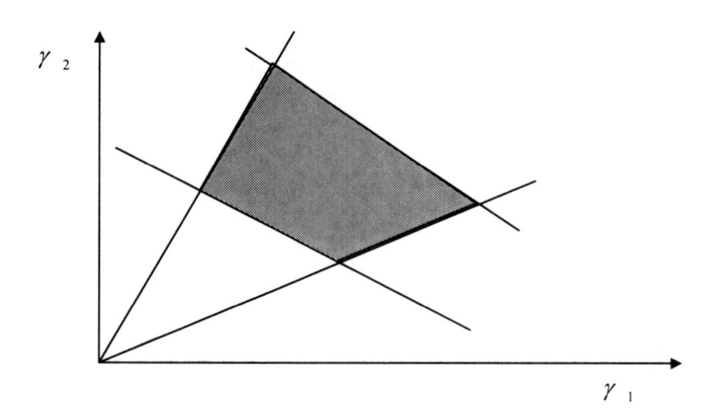

Figure 7: The bounded area is confined more tightly due to the second Type 2 question

In order to find a range for η we have to quantify the hazard rate at a range of covariates and age. Assume P_l and P_u are lower and upper bounds for the probability of failure in a short interval $[t_0, t_0 + \Delta t_0]$ after time t_0, and with covariates value $Z(t_0)$. Then

$$P_l \le h(t_0, Z(t_0))\Delta t_0 \le P_u \tag{9}$$

$$\left(\frac{\beta(t_0)^{\beta-1} e^{\gamma_1 Z_1(t_0)+...+\gamma_m Z_m(t_0)} \Delta t_0}{P_u}\right)^{1/\beta} \le \eta \le \left(\frac{\beta(t_0)^{\beta-1} e^{\gamma_1 Z_1(t_0)+...+\gamma_m Z_m(t_0)} \Delta t_0}{P_l}\right)^{1/\beta},$$

If we define $A = \ln(\frac{\eta^{\beta}}{\beta})$, the proportional hazards model can be depicted as:

$$h(t; Z(t)) = t^{\beta-1} \exp(\gamma_1 Z_1(t) + ... + \gamma_m Z_m(t) - A)$$

This re-parameterization transforms the previous inequality for the value of η to the same format as formula (3), or:

$$\ln P_l \le (\beta - 1)\ln t_0 + \gamma_1 Z_1(t_0) + ... + \gamma_m Z_m(t_0) - A + \ln \Delta t_o \le \ln P_u \tag{10}$$

which means that all inequalities are now a linear function of the unknown parameters.

The question still in need of answering is how these detailed case comparisons are designed. To do so, one might turn to current methods in the field of design of experiments (Berger & Maurer, 2002; Kuehl, 2000; Montgomery, 2005). Methods in the design of experiments (DOE) help one analyze the relationship between some independent variables, say $X = (x_1, x_2,..., x_k)$, and some dependent variables, say $Y = (y_1, y_2,..., y_p)$. In one of its simplest formats we use only one independent variable and the relationship between dependent and independent variables are defined linearly. A general linear model is in the form of:

$$Y = c_0 + c_1 x_1 + c_2 x_2 + ... + c_k x_k + e$$

where e is a variable measuring the error, it usually is assumed to be normally distributed with the mean value of zero.

To frame our problem based on the above model, we define the dependent and independent variables as follows:

$$Y = \frac{h(t_B, Z_B(t_B))}{h(t_A, Z_A(t_B))} = (\beta - 1)\ln(\frac{t_B}{t_A}) + \gamma'(Z_B(t_B) - Z_A(t_A))$$

$$X = [\ln(\frac{t_B}{t_A}), (Z_{B1}(t_B) - Z_{A1}(t_A)), (Z_{B2}(t_B) - Z_{A2}(t_A)), ..., (Z_{Bm}(t_B) - Z_{Am}(t_A))]$$

$$c_0 = 0, c_1 = (\beta - 1), c_2 = \gamma_1, ..., c_{m+1} = \gamma_m$$

Since we have assumed there is a linear relationship between independent and dependent variables and also $c_0 = 0$, we need at least m+1 experiments to be able to estimate values of all parameters. DOE also provides some recommendations on choosing levels of variables and their combinations. Since our model is a linear model and given the other considerations explained in section 8.3.1, we suggest the methodology described in the next section be used when building more detailed questions.

8.3.1 Instructions on How to Design More Detailed and Case Comparison Questions

Care is needed to design the types of questions described in section 8.3. The cases that describe a machine's condition have to make sense to the expert otherwise the validity of his/her answers would be low. In reality there usually are some correlations between the condition indicators (covariates), and also between condition indicators and the working age of the system. Rarely is a group of the condition indicators of a failure process in their normal zones while, at the same time, other condition indicators of the same failure process are in their critical and alarm zones. The condition indicators more or less move together when the condition of a machine changes, however some can have delays, or be less significant indicators compared to others. For example, in the case study discussed at the end of this book there are three covariates: pressure of the 2nd-stage discharge gas (P2); temperature of the 2nd-stage discharge gas (T2); and temperature of the 3rd-stage discharge gas (T3). The correlation between these variables and their correlation with working the age of the component of interest (3rd-stage piston ring) for a sample data history is shown in Table 2.

Table 2

Correlations between Different Covariates and Working Age in a Real Industrial Case

	Age	**P2**	**T2**	**T3**
Age	1			
P2	0.445783	1		
T2	0.516995	0.42163	1	
T3	0.113869	0.52514	0.438398	1

The second example of such correlations is the correlations between different metal particles in circulating oil of diesel engines. The correlation between copper (Cu) and iron (Fe) in the circulating oil of some diesel engines whose failure was believed to be predicted by the behavior of these condition indicators is shown in Table 3.

Table 3

Correlation between Level of Iron and Copper in Circulating Oil of a Diesel Engine

	Cu	**Fe**
Cu	1	
Fe	0.460883	1

The third case relates to condition monitoring of some bearings based on vibration analysis (table 4). The covariates are vibration intensities in different frequency domains. For instance RF14H is defined as the amplitude of the vibration at the frequency equal to one RPM (revolutions per minute). In all the cases demonstrated here, there are relatively high correlations between condition indicators.

Table 4

Correlation between Vibration Intensities in Different Frequency Bounds Related to Failure Process of a Bearing

	RF14H	**RF24H**	**RF54H**
RF14H	1		
RF24H	0.556468	1	
RF54H	0.440146	0.263978	1

We recommend that the values of the covariates in case comparison questions reflect these correlations. One covariate can not be well below its warning limits while another covariate has passed its critical limit. As a general rule we propose that between the values of different covariates describing each case, the following relationship should hold very roughly:

$$\frac{Z_i(t) - L_{Wi}}{L_{ci} - L_{Wi}} \approx \frac{Z_j(t) - L_{Wj}}{L_{cj} - L_{Wj}} \quad \text{for i, j} = 1, 2, \ldots, m$$

where $Z_i(t)$ is ith time-dependent covariate, L_{Wi} and L_{ci} are warning and critical limits of $Z_i(t)$ respectively.

These limits can be replaced by other limits such as emergency limits or alarm limits. This criterion implies that messages that values of different covariates carry in each case should be balanced. The degree to which this criterion should be followed when designing the questions depends on many factors such as correlations between covariates, definitions of the warning and critical limits, and the opinion of the expert. For instance, the expert in our industrial case study gave the following comments for the cases shown below:

2nd-stage discharge gas temperature (F)	310
3rd-stage discharge gas temperature (F)	320
2nd-stage discharge gas pressure (psi)	145
Age (months)	4.5

"I would have expected the 3rd stage gas temperature to be higher based on the 2nd stage pressure you have provided and the 2nd stage temperature. The 3rd stage gas temperature with these other two indicator values should have been up a little closer to 330 or so."

2nd-stage discharge gas temperature (F)	310
3rd-stage discharge gas temperature (F)	330
2nd-stage discharge gas pressure (psi)	140
Age (months)	4.5

"I would expect the 2nd stage pressure to be a little higher based on the 3rd stage temperature of 330 deg."

The expert gave an unexpected answer when he was asked to compare probability of failure for two cases for which the relationship between age and values of the covariates was not normal based on his experience:

	Case A	**Case B**
2nd-stage discharge gas temperature (F)	310	310
3rd-stage discharge gas temperature (F)	325	325
2nd-stage discharge gas pressure (psi)	142	142
Age (months)	4	4.5

"...There is a definite relationship which is not shown in these two cases. 2nd stage pressure and temperatures should be lower in Case A because they have not been in service long."

	Case A	Case B
2nd-stage discharge gas temperature (F)	310	310
3rd-stage discharge gas temperature (F)	325	325
2nd-stage discharge gas pressure (psi)	142	142
Age (months)	3.5	4

"Case A is probably at a higher risk of failure as the 2nd-stage pressure is in alarm condition even though the rings are younger in age."

When the above case was modified to become more realistic the expert's answers changed as below:

	Case A	Case B
2nd-stage discharge gas temperature (F)	310	310
3rd-stage discharge gas temperature (F)	325	325
2nd-stage discharge gas pressure (psi)	142	142
Age (months)	4.5	5

Answer: B is between 1 and 2 times riskier than A.

"Both Case A and B are showing signs of failure. Case B has a slightly higher risk of failure based on age."

To read more about these covariates and their warning, critical, and alarm limits and also their definition see section 11.3.

Based on our experience we suggest that the list of case comparison questions should be shown to the expert after they have been designed to make sure that all the cases make sense to him.

For the cases where age and covariates are strongly correlated different values for age create some behavioral expectations in the expert, and the expert interprets the values of the covariates in light of these expectations. Therefore we suggest that for measuring the effect of age (value of β), it is better not to change values of covariates in two cases that are presented to the expert for comparison (see the last three cases shown above).

Comparing two different cases in terms of probability of a failure (paired comparison) or asking the expert for the ratio of the probabilities of a failure in two cases (rating or ranking) should not be a problem. This statement is based both on the literature (Bock & Jones, 1968; Gescheider, 1997; Kabus, 1976; Murphy & Winkler, 1977; Seaver & Stillwell, 1983; Stillwell, et al., 1982; Torgerson, 1958), and our experience during several experiments and one industrial case study. However since expressing very small probabilities with adequate precision is difficult we suggest that in all case comparison questions, values of age and covariates be close to their warning limits or even critical limits (depending on the definition of each). Also experts pay special attention to the behavior of the covariates and the system when the values of the covariates are high and the system is in the vicinity of failure. This helps experts to estimate those probabilities with more accuracy.

In designing *Type 2* questions, the elicitor should assign values to the age and covariates such that the ratios of the probabilities stay between 1 and 5. If the value of such a ratio exceeds 5, the elicitor can reduce the condition indicators' values for the second case (Case B) to allow the ratio to stay in the suggested range. This will reduce the estimation error since the cases are more comparable for the expert when their probability of occurrence is not very different.

Finding P_l and P_u can be a tough task. Since failures normally do not happen when the system's health conditions are good, it is difficult to quantify the hazard rate in these situations. As mentioned earlier, human beings find expressing very small probabilities with sufficient precision difficult. This suggests that we look for the cases with the highest hazard rates. For example, when the covariates reach their emergency or alarm limits the experts usually suggest that the system be shut down for repair immediately. If you ask them what is likely to happen if they do not stop for repair you will be able to extract information on some probabilities of failure from their answers. The expert can usually be asked directly about the probability of failure. P_l and P_u also can be measured indirectly. For example, an expert might say that if the component is not changed it will fail within two to three days for sure. This means that remaining useful life (RUL), also known as the *mean residual life*, is between two and three days in most of the cases (depending on the definition of "for sure" by that expert) for that component in that situation. Using this answer and some mathematics we can find lower and upper bounds for the hazard rate function as shown below. More detailed explanations of derivations of RUL for a PHM with a time-dependent covariate are given in Banjevic and Jardine (2006).

$$RUL(t)=e(t)=E(T-t \mid T>t) \qquad \textbf{(11)}$$

where t is current age and T is a variable describing the time to failure.

In the case of a PHM with time-dependent covariates the above formula looks like this:

$$RUL(t, Z(t)) = e(t, Z(t)) = E(T-t \mid T>t, Z(t))$$

After finding the reliability function ($R(.|t, Z(t))$), we have:

$$e(t, Z(t))= \int_t^{\infty} R(s \mid t, Z(t))ds = \int_t^{\infty} e^{h_0(s)e^{\gamma Z(s)}} ds = \int_t^{\infty} \exp(\frac{\beta}{\eta}\left(\frac{s}{\eta}\right)^{\beta-1} e^{\gamma_1 Z_1(s)+...+\gamma_m Z_m(s)})ds \qquad \textbf{(12)}$$

If (t, Z(t)) describes a situation very close to failure, then one can assume that values of t and Z(t) remain constant until the component fails. Using this approximation we have:

$$e(t, Z(t)) \approx \frac{1}{h(t, Z(t))} = \frac{1}{\frac{\beta}{\eta}\left(\frac{t}{\eta}\right)^{\beta-1} \exp(\gamma_1 Z_1(t) + ... + \gamma_m Z_m(t))}$$

$$= \frac{1}{t^{\beta-1} \exp(\gamma_1 Z_1(t) + ... + \gamma_m Z_m(t) - A)} \qquad \textbf{(13)}$$

For instance if the expert believes that the a component currently having age of t and covariates level of Z(t) will fail between two and three days with 90% confidence, it implies that the following statements are true 90% of time:

$$2 \le e(t, Z(t)) \le 3,$$

$$2 \le \frac{1}{t^{\beta-1} \exp(\gamma_1 Z_1(t) + ... + \gamma_m Z_m(t) - A)} \le 3,$$

$$\frac{1}{3} \le t^{\beta-1} \exp(\gamma_1 Z_1(t) + ... + \gamma_m Z_m(t) - A) \le \frac{1}{2}$$

This provides a complete set of inequalities needed for restricting all the parameters of PHM.

Note: the above assumption creates some bias toward overestimation of the hazard because it assumes the value of the hazard will remain constant until failure. For cases where failure is imminent, this approximation might not be bad, but we should take into account this deficiency while building inequalities. More can be done to reduce this bias. For example one might ask the expert to provide the values of the condition indicators at the predicted failure so that the elicitor can average values of both hazards and use them in formula 13.

8.3.2 Required Number of Inequalities

The number of required inequalities depends on a few factors such as:

1. *Number of parameters*: the more parameters there are, the more inequalities are needed to effectively bound the feasible area. Mathematically speaking, to bound n variables, at least n + 1 inequalities are needed. This is consistent with what DOE (design of experiments) suggests.
2. *Level of uncertainty in experts' answers*: Since most of the inequalities are not active all the time, due to the uncertainties in experts' answers, we definitely need more than suggested by the previous formula. (Chapter 9 shows how different cases with different levels of uncertainty should be handled.)
3. *Computing facilities limitations*: more inequalities means that the simulating software requires more time to find the required number of samples, therefore this works as a practical boundary.

Considering what has been discussed so far and including our experience, we suggest the following number of questions for each combination of the parameters:

For each covariate coefficient, two *Type 2* questions. This implies that for each coefficient of covariates we should design two *Type 2* case comparison questions in which the covariate value of only one parameter is different and other parameters are constant. In this way we can calculate an upper and a lower bound for each parameter separately. We suggest two such questions in order to increase the reliability of the answers. For example in the Dofasco case study we asked the expert to compare the following pairs of cases and tell us how many times Case B is at higher risk of failure compared to Case A:

	Case A	**Case B**
2nd-stage discharge gas temperature (F)	310	310
3rd-stage discharge gas temperature (F)	330	330
2nd-stage discharge gas pressure (psi)	142	147
Age (months)	4	4

Answer: B is between 1.5 and 2.5 times riskier than A.

Temperatures are the same in both cases but 2nd-stage pressure in Case B would indicate a higher probability of failure.

	Case A	**Case B**
2nd-stage discharge gas temperature (F)	310	310
3rd-stage discharge gas temperature (F)	330	330
2nd-stage discharge gas pressure (psi)	142	145
Age (months)	4	4

Answer: B is between 1 and 2 times riskier than A.

Based on the first answer, the average value of the coefficient of 2nd stage discharge gas is estimated to be $\frac{\ln(2)}{5} = 0.1386$ and based on the second answer its average value is estimated to be $\frac{\ln(1.5)}{3} = 0.1351$. As shown, these two estimates are very close to each other.

The first answer also suggest that the following upper and lower limits for the coefficient of the 2nd stage discharge gas pressure holds 90% of the time:

Lower limit = 0.0811

Upper limit = 0.1832

However, the second question suggests the following as the lower and upper limits:

Lower limit = 0.00

Upper limit = 0.2310

For each pair of coefficients of the covariates, two *Type 2* and two *Type 1* questions.

For each group of three and more of coefficients of the covariates, two *Type 2* questions.

For instance in the Dofasco case study we have:

	Case A	**Case B**
2nd-stage discharge gas temperature (F)	300	305
3rd-stage discharge gas temperature (F)	315	320
2nd-stage discharge gas pressure (psi)	135	140
Age (months)	4	4

Answer: B is between 2 and 3 times riskier than A.

	Case A	**Case B**
2nd-stage discharge gas temperature (F)	305	310
3rd-stage discharge gas temperature (F)	325	330
2nd-stage discharge gas pressure (psi)	140	145
Age (months)	4	4

Answer: B is between 2 and 3.5 times riskier than A.

	Case A	**Case B**
2nd-stage discharge gas temperature (F)	300	305
3rd-stage discharge gas temperature (F)	320	325
2nd-stage discharge gas pressure (psi)	135	140
Age (months)	4	4

Answer: B is between 2 and 3 times riskier than A.

Answers to the first and third questions imply $0.1386 \le \gamma_1 + \gamma_2 + \gamma_3 \le 0.2197$ is true 90% of the time

The answer to the second question is equivalent to $0.1386 \le \gamma_1 + \gamma_2 + \gamma_3 \le 0.2505$ being true 90% of the time. Here assume γ_1, γ_2, and γ_3 are the coefficients of 2nd-stage discharge gas temperature, 3rd-stage discharge gas temperature, and 2nd-stage discharge gas pressure respectively.

Based on our experience, the questions in which all the covariates change provide the most accurate estimates. Because of the correlation between different covariates it is more realistic (and makes more sense to the expert) that when one of the covariates shifts the others shift as well.

- Five *Type 2* questions for measuring the age
- Five questions to measure the upper and lower bound of the hazard

Based on what has been suggested when there are three covariates we need around 3*2 + 2*(3*2) + 2 + 5 + 5 = 25 case comparison questions. This number can be increase if we receive contradictory answers from the expert.

9. BUILDING PRIOR DISTRIBUTIONS

To build the prior distribution of the parameters we can sample from the feasible parameter space created by the bounding inequalities obtained during expert knowledge gathering. To randomly select a point $(\beta, A, \gamma_1, ..., \gamma_m)$ in the space restricted by the bounding inequalities we take the following steps:

1. Find the upper and lower limits for all the parameters either by using optimization software (e.g. maximize β given the constraints) or by trial and error.

$$\beta_l \leq \beta \leq \beta_u, A_l \leq A \leq A_u, \gamma_{il} \leq \gamma_i \leq \gamma_{iu}, \quad 1 \leq i \leq m.$$

2. Generate m+2 random numbers (denoted as $\beta^1, A^1, \gamma_1^1, ... \gamma_m^1$) using the uniform distribution function for each parameter.

If $(\beta^1, A^1, \gamma_1^1, ... \gamma_m^1)$ satisfies all bounding inequalities we consider it as one realization of the parameters, or a sample from the joint prior distribution function of the parameters. The procedure is repeated until a specified number of samples (say n) is obtained. This can be a lengthy process especially if the number of parameters exceeds five. To handle this problem one can use conditional sampling. This means that the range of each parameter will be a function of the values already generated for the other parameters. In this sense it is similar to the Gibbs sampling method (Brooks et al., 1998). For example, after generating a random number for β using *Uniform*(β_l, β_u), we can redefine upper and lower limits for A using the current value of β. This will reduce the range of values for A which results in avoiding values that do not satisfy the constraints. This in turn requires more computation to find the new upper and lower limits for the parameters. Therefore this method is suggested only when the feasible space is very crooked and twisted since for other cases enough samples can usually be obtained in a reasonably short time. In the case study presented later, condition sampling was used for the last of its five parameters (A in this case), and it obtained around 5,400 samples every 15 seconds.

It also is possible to take into account experts' uncertainties regarding their answers. To do so, the inequality obtained with a degree of confidence $p \times 100\%$, $0<p<1$, is used (is active) $p \times 100\%$ of times we sample from the feasible space. So, if an expert is 80% confident about the answer to a question, the inequality obtained from that answer would be active with a probability of 80% when we sample from the feasible space. Actually, using this method we give different weights to the different regions of the feasible space, which makes our prior distribution more flexible in accommodating the experts' opinion (see Figures 8–12).

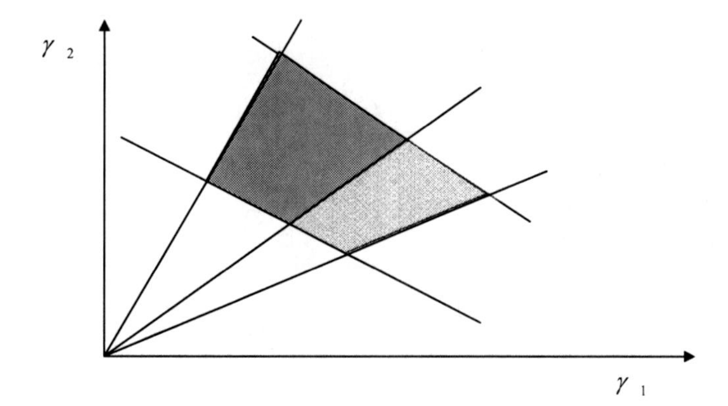

Figure 8: Inequalities that are not active all the time during the sampling process give different weights to different parts of the parameters' space. See also Figures 9, 10, and 11

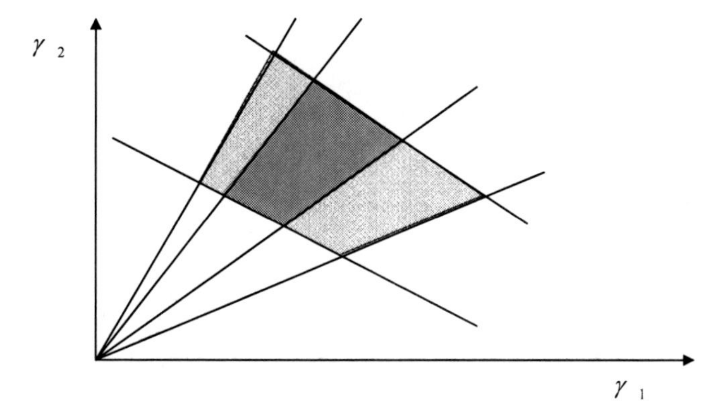

Figure 9: Further bounding inequalities

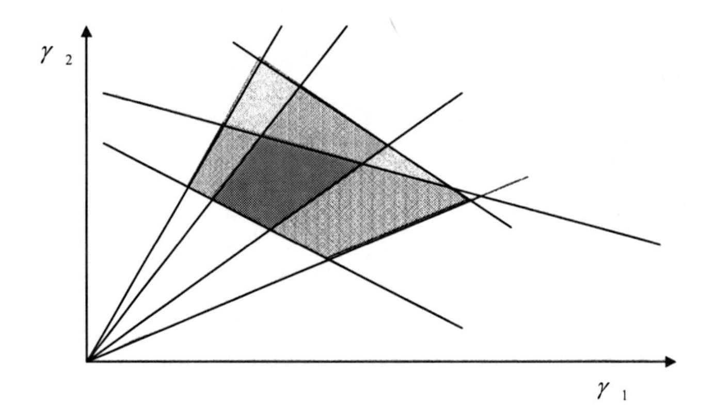

Figure 10: Further bounding inequalities

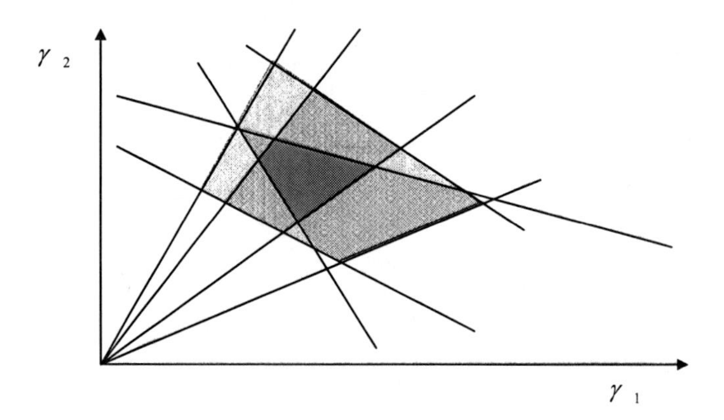

Figure 11: Further bounding inequalities

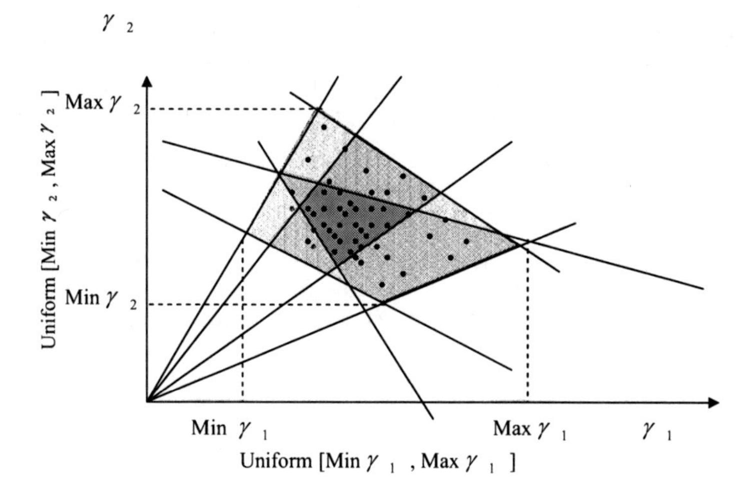

Figure 12: Sampling from the feasible space created by the inequalities

9.1 Representing Prior Distributions

Now that we have samples of prior distribution, its moments can be used in describing the behavior of the distribution and also in decision-making as presented in sections 4.4 and 9.9. For example we can use the first moment as in the PHM formula to measure the hazard rate of a system. The first and the second moments of the prior distribution can be simply calculated using the following formulas:

$$\hat{\beta} = \frac{1}{n}\sum_{i=1}^{n}\beta^i, \qquad s.d.(\beta) = \sqrt{\frac{1}{n-1}\sum_{i=1}^{n}(\beta^i - \hat{\beta})^2}$$

$$\hat{A} = \frac{1}{n}\sum_{i=1}^{n}A^i, \qquad s.d.(A) = \sqrt{\frac{1}{n-1}\sum_{i=1}^{n}(A^i - \hat{A})^2} \qquad (14)$$

$$\cdots\cdots \qquad \cdots\cdots$$

$$\cdots\cdots \qquad \cdots\cdots$$

$$\hat{\gamma}_m = \frac{1}{n}\sum_{i=1}^{n}\gamma^i{}_m \qquad s.d.(\gamma_m) = \sqrt{\frac{1}{n-1}\sum_{i=1}^{n}(\gamma^i{}_m - \hat{\gamma}_m)^2}$$

These samples can also define a non-parametric distribution for the prior distribution as explained shortly. Common methods (Gelman et al., 2004) suggest working with closed form prior distributions which then are updated to posterior distributions using Bayes' rule. If the posterior distribution is in the form of known distributions it would be easy to obtain its moments and use them as updated estimations for the values of the parameters. If the posterior distribution is complicated, numerical methods can be applied to sample from the posterior distributions. These samples are then used in estimating the values of the parameters. The approach in this research is different, however. Knowing that the posterior distribution of the parameters of time varying PHM is too complicated, and also that our methodology directly obtains samples from prior distributions, we use these samples in the updating process directly. Using the traditional method we would have to choose a closed-form multidimensional distribution function to approximate the sample data obtained from the prior distribution. Then applying Bayes' rule, there would be considerable calculation required to get the posterior distribution. Afterward we could use MCMC to obtain samples from the posterior distribution. This procedure involves too much approximation, and we expect that our proposed methodology would result in a more accurate outcome.

Note: There are high correlations between sampled values of the parameters. For instance, in some cases the correlation between two regression coefficients (γ_i and γ_j) in practice can be as high as 0.8 or even more. In the Dofasco case study we obtained the following correlation matrix for the parameters.

Table 5

Correlation between Parameters of PHM Obtained Using Expert Knowledge

	β	A	γ_1	γ_2	γ_3
β	1	**0.2412**	-0.0007	0.0091	0.0038
A	**0.2412**	1	0.0200	-0.0413	**0.3380**
γ_1	-0.0007	0.0200	1	**-0.2202**	**-0.7429**
γ_2	0.0091	-0.0413	**-0.2202**	1	**-0.3938**
γ_3	0.0038	**0.3380**	**-0.7429**	**-0.3938**	1

This would make the job of those who choose to follow the traditional methods much more difficult because in the prior and posterior distribution function they have to include the correlations between all the parameters. This complicates both the job of finding a posterior distribution and that of sampling from it. To simplify the problem, some authors (Faraggi & Simon, 1997) make the assumption that there is no correlation between the parameters. Based on the results of the experiments, This assumption in our opinion is unrealistic based on the results of the experiments.

To present the prior distribution of the parameters, a notation similar to a multidimensional histogram is used. First we divide the range of each parameter into a number of equal intervals. Let the range of β, $[\beta_l, \beta_u]$, be divided into d_1 equal intervals, which can be shown as:

$$[\beta_l + K\Delta_1, \beta_l + (K+1)\Delta_1),$$

$$\Delta_1 = \frac{\beta_u - \beta_l}{d_1},$$

$$K = 0, 1, \ldots, \mathrm{d}_1 - 1.$$

Let the range of A be divided into d_2 equal intervals and the range of γ_i be divided into d_{i+2} equal intervals. In this way the space containing the feasible solution is divided into $Q = d_1 \times d_2 \times ... d_{m+2}$ equal "blocks." We assume that the value of joint prior distribution, which is denoted by $f_{\beta,A,\gamma_1,...,\gamma_m}(.)$ is constant inside each block. We may increase the number of blocks to obtain the required accuracy using the previous method. The following intervals

$$\beta \in [\beta_l + (i-1) \times \Delta_1, \beta_l + i \times \Delta_1),$$

$$A \in [A_l + (j-1) \times \Delta_2, A_l + j \times \Delta_2),$$

$$\gamma_1 \in [\gamma_{1l} + (k-1) \times \Delta_3, \gamma_{1l} + k \times \Delta_3),$$

.

.

$$\gamma_m \in [\gamma_{ml} + (r-1) \times \Delta_{m+2}, \gamma_{ml} + r \times \Delta_{m+2})$$

define the block $b_{i,j,k,...,r}$ with volume $V = \prod_{i=1}^{m+2} \Delta_i$, and do the same for all blocks.

We then estimate the prior distribution by:

$$f_{\beta,A,\gamma_1,...,\gamma_m}(\beta, A, \gamma_1, ..., \gamma_m) \approx f^{(n)}{}_{\beta,A,\gamma_1,...,\gamma_m}(\beta, A, \gamma_1, ..., \gamma_m)$$

$$= \frac{\text{Total samples inside } b_{i,j,k,...,r} + 1}{(\mathrm{n} + Q) \times V} \qquad \textbf{(15)}$$

$$\beta, A, \gamma_1, ..., \gamma_m \in b_{i,j,k,...,r}$$

where n is the number of samples and Q is the total number of blocks.

In the numerator, 1 is added to ensure that all blocks have positive density. If a block has zero value for the density function, its posterior would become zero regardless of the value of the likelihood, which we want to avoid. To compensate for adding 1 to each block, Q is added to the denominator to make sure hat the sum of probabilities over all blocks is equal to 1. We have also multiplied the denominator by the volume of the block to create the density function. Adding one sample to each block can significantly affect the prior distribution if total sample size is small. To avoid this, we choose $\mathrm{n} > 5 \times \mathrm{Q}$. "Thus, we have defined the prior distribution function and estimated it using sample data (see Figure 13). The next level would be to find the posterior distribution.

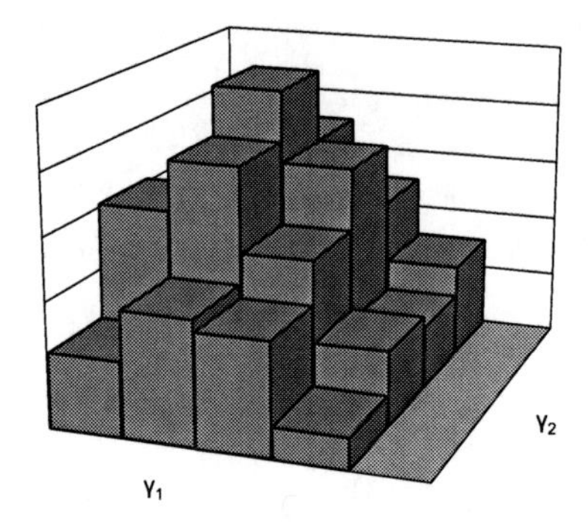

Figure 13: Samples from the feasible space will form a multidimensional histogram

9.2 Required Characteristics of Prior Distributions

The literature suggests the following characteristics for a suitable prior distribution (Gelman et al., 2004). After explaining each characteristic we show how our prior distribution can satisfy these criteria.

9.2.1 Capable of Adopting Expert's Knowledge

It is best if the parameters of prior distributions (known as hyperparameters) make sense to the experts. In this case experts will be able to relate hyperparameters to their knowledge. In this research we do not have parametric prior distribution, rather the prior distribution is obtained using samples taken from the feasible space. The feasible space itself is defined based on expert knowledge. Therefore we believe that our methodology satisfies this criterion. For example if an expert believes in the following statement regarding two cases, A and B:

$$b_l \leq \frac{h(t_B, Z_B(t_B))}{h(t_A, Z_A(t_B))} \leq b_u,$$

or equivalently,

$$\ln b_l \leq (\beta - 1)\ln(\frac{t_B}{t_A}) + \gamma'(Z_B(t_B) - Z_A(t_A)) \leq \ln b_u,$$

$$\ln b_l \leq (\beta - 1)\ln(\frac{t_B}{t_A}) + \gamma_1(Z_{1B}(t_B) - Z_{1A}(t_A)) + ... + \gamma_m(Z_{mB}(t_B) - Z_{mA}(t_A)) \leq \ln b_u$$

All the samples randomly generated then have to satisfy this criterion based on our methodology:

$$\ln b_l \le (\beta^1 - 1)\ln(\frac{t_B}{t_A}) + \gamma^1{}_1(Z_{1B}(t_B) - Z_{1A}(t_A) + ... + \gamma^1{}_m(Z_{mB}(t_B) - Z_{mA}(t_A)) \le \ln b_u$$

$$\ln b_l \le (\beta^2 - 1)\ln(\frac{t_B}{t_A}) + \gamma^2{}_1(Z_{1B}(t_B) - Z_{1A}(t_A) + ... + \gamma^2{}_m(Z_{mB}(t_B) - Z_{mA}(t_A)) \le \ln b_u$$

....

....

$$\ln b_l \le (\beta^n - 1)\ln(\frac{t_B}{t_A}) + \gamma^n{}_1(Z_{1B}(t_B) - Z_{1A}(t_A) + ... + \gamma^n{}_m(Z_{mB}(t_B) - Z_{mA}(t_A)) \le \ln b_u$$

Resulting in:

$$\ln b_l \le \frac{\sum_{i=1}^{n}(\beta^i - 1)\ln(\frac{t_B}{t_A}) + \gamma^i{}_1(Z_{1B}(t_B) - Z_{1A}(t_A)) + ... + \gamma^i{}_m(Z_{mB}(t_B) - Z_{mA}(t_A))}{n} \le \ln b_u$$

or

$$b_l \le \sqrt[n]{\prod_{i=1}^{n}(\frac{t_B}{t_A})^{(\beta^i - 1)} \exp(\gamma^i{}_1(Z_{1B}(t_B) - Z_{1A}(t_A)) + ... + \gamma^i{}_m(Z_{mB}(t_B) - Z_{mA}(t_A)))} \le b_u$$

$$b_l \le (\frac{t_B}{t_A})^{\left(\frac{1}{n}\sum_{i=1}^{n}(\beta^i - 1)\right)} \exp(\left(\frac{1}{n}\sum_{i=1}^{n}\gamma^i{}_1\right)(Z_{1B}(t_B) - Z_{1A}(t_A)) + ... + \left(\frac{1}{n}\sum_{i=1}^{n}\gamma^i{}_m\right)(Z_{mB}(t_B) - Z_{mA}(t_A))) \le b_u$$

$$b_l \le (\frac{t_B}{t_A})^{\hat{\beta}-1} \exp(\hat{\gamma}_1(Z_{1B}(t_B) - Z_{1A}(t_A)) + ... + \hat{\gamma}_m(Z_{mB}(t_B) - Z_{mA}(t_A))) \le b_u$$

Similar statements can be derived for the case where upper and lower bounds are defined for the hazard using equation 9 in Chapter 8.

This shows that any knowledge transferred through the inequalities will be reflected in the mean of the sampled data from the feasible space.

9.2.2 Acting as Non-informative Prior Distribution If No Expert Judgment Is Available

In such a case the impact of data for estimating parameters will be dominant. For instance, an expert might have some knowledge about one of the covariates but no knowledge about the other one. Hence the process must be able to incorporate the expert knowledge regarding the first covariate and build a non-informative prior distribution for the second one.

In our methodology if an expert has no information about the effect of a parameter, that parameter will not be bounded with inequalities; it will have a uniform distribution which is a non-informative prior distribution. The significance of that parameter can be measured after posterior distribution has been built.

9.2.3 Facilitating the Updating Process

There are many methods for choosing the best prior distribution in different situations (Group Paper, 1998; Kadane & Wolfson, 1998; O'Hagan, 1998; Oien, 1998; Percy, 2004; Singpurwalla & Song, 1988; Walls & Quigley, 2001). The preference is to have a conjugate prior distribution which means that both prior and posterior distributions are in the same family of distribution functions so that for each updating process the same procedure can be used. It also helps result in a simpler posterior distribution. However in this case (PHM with time-dependent covariate) there is no parametric conjugate prior distribution.

If we use the above definition for conjugate distribution, we can say that our non-parametric prior distribution is a conjugate prior distribution since the updating process is the same every time we collect data and the posterior distribution has the same format as for prior distribution.

9.2.4 Capable of Combining Knowledge from Different Sources

Consider the case where more than one expert is available. Whenever the decision required is important we might want to exploit all available expertise. We can incorporate the knowledge of different experts at two levels. It can be done during the elicitation process or the knowledge of each expert can be elicited separately and the prior distributions combined subsequently. Below the two methods are explained and some related background information provided.

9.2.4.1 Combining Experts' Knowledge While Eliciting Opinions

As mentioned earlier, knowledge of experts can be combined during the elicitation process. In this case the experts interact as a group. There are two main approaches for handling this situation. One is to bring the experts together to discuss unknown quantities of interest. Their aggregated knowledge is then used in building prior distributions. In effect we have considered the group as a single individual. This is called the *behavioral aggregation* approach (Philips, 1999 (see Garthwaite et al., 2005)). However when a group of people come together there may well be psychological issues affecting their interaction. These issues require a facilitator with strong knowledge and experience to lead the group discussions and elicitation process. One problem is that some experts might censor themselves in favor of higher ranking and more dominant experts. There is also the possibility of overlapping experiences being overweighted because they have been repeated many times during group discussions. In some situations the group might suppress dissenting views to reach consensus. On the other hand, there can be occasions when the group would not reach consensus at all.

All of these drawbacks have caused some practitioners use an alternative method called *Delphi* (Adler & Ziglio, 1996; Bowles, 1999; Pill, 1970), a formal method for managing interactions during knowledge elicitations in a group. In this method the knowledge of each expert is extracted separately. Then the elicited knowledge is sent anonymously to the other members of the group. In addition to individuals' opinions, each expert is required to explain how he/she has arrived at those conclusions. Other members of the group review the opinions and the accompanying reasoning behind them Then each expert is invited to revise his/her opinion after observing other member's opinions and explanations. This is an iterative process that continues until all the members agree on the elicited knowledge. Although this methodology removes some of the concerns raised by the behavioral aggregation approach, it can be a less efficient way of sharing of knowledge especially when the group is large.

In some fields, such as medical science, you can usually find a few experts on a topic. For example Chaloner and Rhame (2001) elicited opinions from 58 HIV experts. When there are so many experts, any of the methods described earlier or others not discussed mentioned here can be used. However in the field of reliability it is uncommon to see many experts on a subject particularly if we are interested in elicitation of the failure process of complex machines. We might be able to find one, two, or in some cases up to five people with enough knowledge regarding a sophisticated piece of equipment. In this book we emphasize the word "sophisticated" because companies apply condition-based maintenance techniques to their "sophisticated" equipment. This means that in our frame (knowledge about the parameters of PHM) we should not expect to see more than four or five experts for each unit of complex equipment. When we have only one, two, or even three experts it is not feasible to use the Delphi method because the experts will know who provided which opinion. In such case we suggest using the *behavioral aggregation* method under the supervision of a highly skilled facilitator capable of managing the meetings so as to ensure all opinions are heard and given appropriate considerations. We might want to use Delphi method if we have more than four or five experts. In either case, the experts would be presented with the cases explained in section 8.3. Then, using the behavioral aggregation method, we would invite them to give their answers and discuss it within the group. When they have reached an agreement on each case we would move on to the next one. The results would be a set of inequalities based on the aggregated knowledge of the group. Random numbers which satisfy these inequalities are considered to be random samples of the prior distributions based on the combined knowledge of the group.

9.2.4.2 Combining Separate Elicitations

Some approaches have been suggested above for eliciting and blending the knowledge of different experts. In this section we want to review methodologies that can be used to combine prior distributions obtained based on the knowledge of different experts. Several authors have reviewed formal methods of combining probability distributions. Genest and Zidek (1986) provide a very useful bibliography on the topic. French (1985) has also provided an extensive review on different techniques applied in attaining group consensus regarding probability distributions.

There are two well-known methodologies which both fall in a category known as *opinion pools*. The linear opinion pool is a weighted average of different individuals' probability distributions. The logarithmic opinion pool is a normalized weighted geometric mean. This is equivalent to applying a linear pool to the logarithms of the individual probability densities and then normalizing the results (Garthwaite, 2005).

It is an important characteristic of an opinion pool that it be externally Bayesian (Genest & Zidek, 1986). This means that the posterior distribution of the opinion pool of different prior distributions should be equal to the opinion pool of the posterior distributions of the constituting priors. Based on this definition, the linear opinion pool fails to meet this criterion. However when the weights (given to different experts) sum up to one the logarithmic pool has this property. The second important property is *invariance to event combination*. Assume there are two exclusive events, A and B, and we define C as the union of A and B (A or B). If we elicit an expert's knowledge regarding the probabilities of A, B, and C, assuming the expert's knowledge is coherent, we expect to see:

$$P(C) = P(A) + P(B)$$

The *combination invariance* property, also known as marginalization criterion, requires that the same relation hold for the pooled probabilities of A, B, and C. McConway (1981) showed that only the linear opinion pool satisfies this criterion. Therefore it is not possible to find a mechanistic opinion pooling method that is externally Bayesian and also satisfies the marginalization criterion (Garthwaite, 2005).

Another drawback of the logarithmic opinion pool is that if the probability of an event based on one expert's view is zero, the probability of the pool for that event also would be zero. To learn more about such issues see Genest & Zidek (1986).

There are cases where the qualifications of the experts are not equal. In the area of reliability some experts have been working with a piece of machinery for 15 years, while others have less experience. Although we do not want to exclude knowledge of those with less experience, giving equal weights to the experts who have different level of expertise does not seem to be appropriate. Fortunately, both linear and logarithmic pools allow the assignment of different weights to the experts so we can assign higher weights to those we believe have more accurate opinions. Cooke (1991) has proposed a methodology that helps facilitators to assign appropriate weights to different experts'opinions. This is done based on the ability of the experts in assessing the distribution for *seed* variables. *Seed* variables are quantities whose true value is known to the facilitator but not to the experts. Cooke & Goossens (2000) has provided some evidence to show that using the suggested procedure for weighted averaging produces better results than giving equal weights to different experts. Another issue with opinion pooling techniques is that they cannot account for double counting opinions (Garthwaite, 2005). For example, if the knowledge of two experts overlaps considerably, we are double counting that particular knowledge when using opinion pooling techniques. Common sense suggests that we should reduce the weight given to those whose knowledge overlaps. However there is no sound methodology for doing that.

In this book we use both linear and logarithmic pooling approaches. For linear pooling there are two choices. We can use empirical prior distributions based on each expert's opinion, or we can use the corresponding sampled parameters values directly. The second method is expected to use fewer approximations than the first. For the logarithmic pooling we would not be concerned about having zero value for any block since we ensure that there is at least one sample in each block. This would prevent the problem mentioned earlier regarding the logarithmic pooling method.

9.2.4.2.1 Using Linear Pools to Combine Prior Distributions of Different Experts

Assume that there are k experts. Based on each expert's answers we define a feasible space. For each parameter we have an upper and lower bound according to each expert's opinion. For every parameter we find the widest range that includes all the values of that parameter based on the knowledge of all the experts. This can be considered as the union of the values of that parameter according to all k experts. Using this procedure we have defined new upper and lower bounds for all parameters. Using the procedure explained in this chapter we now calculate prior distributions based on each expert's opinion. Let $f_p^{(n)}{}_{\beta,A,\gamma_1,\dots,\gamma_m}(\beta,A,\gamma_1,\dots,\gamma_m)$ denote the empirical prior distribution of pth ($1 \le p \le k$) expert based on n sample data obtained from the feasible space. Also let w_p be the weight given to the pth expert. Obviously we have: $\sum_{p=1}^{k} w_p = 1$.

Therefore the empirical prior distribution based on the aggregated knowledge of the k experts would be:

$$f^{(n)}{}_{\beta,A,\gamma_1,\dots,\gamma_m}(\beta,A,\gamma_1,\dots,\gamma_m) = \sum_{p=1}^{k} w_p f_p^{(n)}{}_{\beta,A,\gamma_1,\dots,\gamma_m}(\beta,A,\gamma_1,\dots,\gamma_m)$$

(16)

$$\beta,A,\gamma_1,\dots,\gamma_m \in b_{i,j,k,\dots,r}$$

Note that the same number of samples were used for each expert's prior distribution because we wanted to be consistent in terms of quality of the prior distributions. This is based on the fact that more sampled data from the feasible space means more precision in estimating the prior distributions.

As indicated earlier we also can use the sampled data directly. We sample n_p numbers from the feasible space obtained as a result of the elicitation process of the pth expert such that $w_p = \frac{n_p}{\sum_{i=1}^{k} n_i}$. Then we pool all the sampled data and using the methodology illustrated in section 9.1 we find the empirical prior distribution. This sample can act as a sample of the whole group of experts and therefore the formulas described in section 9.1 can be applied to this one with no change.

9.2.4.2.2 Using Logarithmic Pools to Combine Prior Distributions of Different Experts

We use the notations defined in section 9.1 to show how logarithmic pools can be used to aggregate knowledge of different experts. The value of the prior distribution for each block is the normalized geometric average of values of prior distributions based on all other experts for that block. Mathematically this is:

$$f^{(n)}{}_{\beta,A,\gamma_1,\dots,\gamma_m}(\beta,A,\gamma_1,\dots,\gamma_m) = \frac{\sqrt[\sum_{p=1}^{k} w_p]{\prod_{p=1}^{k}[f_p{}^{(n)}{}_{\beta,A,\gamma_1,\dots,\gamma_m}(\beta,A,\gamma_1,\dots,\gamma_m)]^{w_p}}}{\sum_{\text{for all } b_{i,j,k,\dots,r}} \sqrt[\sum_{p=1}^{k} w_p]{\prod_{p=1}^{k}[f_p{}^{(n)}{}_{\beta,A,\gamma_1,\dots,\gamma_m}(\beta,A,\gamma_1,\dots,\gamma_m)]^{w_p}}} \qquad \textbf{(17)}$$

$$\beta,A,\gamma_1,\dots,\gamma_m \in b_{i,j,k,\dots,r}$$

Since the prior distribution over all blocks is greater than zero, it will remain so for the prior distribution based on the knowledge of k experts.

9.3 Using the Sample Data in Decision-Making

If we do not have any data we can use the sample data of the prior distribution directly in our decision-making. Two different applications of Bayesian statistics were introduced in section 4.4. In this section we illustrate how they can be applied using the sample data. If we want to estimate the parameters we can use sample statistics of the data as illustrated in formula 14. For example each parameter can be estimated by taking the mean value of the corresponding data in the sample. This is called the mean estimator or ME.

$$\hat{\beta} = \int_{\beta,A,\gamma_1,\ldots,\gamma_m} \beta \times f^{(n)}{}_{\beta,A,\gamma_1,\ldots,\gamma_m}((\beta, A, \gamma_1,\ldots,\gamma_m) d\beta \times dA \times d\gamma_1 \times \ldots \times d\gamma_m \approx \frac{1}{n}\sum_{i=1}^{n} \beta^i,$$

$$\hat{A} = \int_{\beta,A,\gamma_1,\ldots,\gamma_m} A \times f^{(n)}{}_{\beta,A,\gamma_1,\ldots,\gamma_m}((\beta, A, \gamma_1,\ldots,\gamma_m) d\beta \times dA \times d\gamma_1 \times \ldots \times d\gamma_m \approx \frac{1}{n}\sum_{i=1}^{n} A^i, \qquad \textbf{(18)}$$

...

We use these estimates to find the value of the decision function $C(x, \beta, A, \ldots \gamma_m)$. The result is: $C(x, \hat{\beta}, \hat{A}, \ldots \hat{\gamma}_m)$. If we want to make better use of the information we have about the parameters, we can estimate the integral introduced in section 4.4 with summation:

$$C(x) = \int_{\beta,A,\gamma_1,\ldots,\gamma_m} C(x, \beta, A, \ldots \gamma_m) \times f^{(n)}{}_{\beta,A,\gamma_1,\ldots,\gamma_m}((\beta, A, \gamma_1,\ldots,\gamma_m) d\beta \times dA \times d\gamma_1 \times \ldots \times d\gamma_m$$

$$\approx \frac{1}{n}\sum_{i=1}^{n} C(x, \beta^i, A^i, \ldots \gamma^i{}_m) \qquad \textbf{(19)}$$

There is another option for estimating the parameters which is similar to the maximum likelihood estimation. It entails finding the maximum value of the prior distribution. We find its corresponding block and the center of that block, or the average of all sampled data in that block can be considered as an estimation of the parameters. This works well if we have one candidate as the highest point in the prior distribution.

10. DESIGNING AN UPDATING PROCESS

A procedure has to be in place to update the prior distributions of the PHM parameters whenever a new set of data arrives. This process must be designed in a way that facilitates the updating process. As explained in section 3.3, in a PHM with time-dependent covariates inspection records are obtained periodically to evaluate the state of the equipment. Data arrives in the following form:

$$(t_i, \delta_i, Z_1(t_i), Z_2(t_i), ...Z_m(t_i)) \qquad \textbf{(20)}$$

where t_i is the current observation time, $Z_1(t_i), Z_2(t_i), ...Z_m(t_i)$ are values of covariates at t_i, and δ_i is a dummy variable which denotes the current state of the history. If t_i is the failure time, δ_i=1; and if t_i is the suspension time, δ_i=0.

When we observe the data set $y = \{(t_i, \delta_i, Z_1(t_i), Z_2(t_i), ...Z_m(t_i))\}$, the posterior distribution based on Bayes' rule is proportional to the likelihood function of the data multiplied by the prior distribution:

The likelihood function is usually denoted by $L(y / \beta, A, ..., \gamma_m)$

If we denote $\theta = (\beta, A, \gamma_1, ..., \gamma_m)$, and the posterior density function by $g_\theta(./ y)$, then we have:

$$g_\theta(\beta, A, ..., \gamma_m / y) \propto f_\theta(\beta, A, ..., \gamma_m) \times L(y / \beta, A, ..., \gamma_m),$$

If we integrate over all the values of θ we have:

$$g_\theta(\beta, A, ..., \gamma_m / y) = \frac{f_\theta(\beta, A, ..., \gamma_m) \times L(y / \beta, A, ..., \gamma_m)}{\int\limits_{\beta, A, ..., \gamma} f_\theta(\beta, A, ..., \gamma_m) \times L(y / \beta, A, ..., \gamma_m) \times d\beta \times dA \times ... \times d\gamma_m}$$

There are several approximations to $g_\theta(\beta, A, ..., \gamma_m / y)$. For the sake of simplicity we will consider the approximation:

$$\hat{g}_\theta(\beta, A, ..., \gamma_m / y) = \frac{f^{(n)}{}_\theta(\beta, A, ..., \gamma_m) \times L(y / \beta_i^*, A_j^*, ..., \gamma_{l\,m}^*)}{\sum\limits_{i,j,..r} \int\limits_{\beta, A, ... \gamma_m \in b_{i,j,...l}} f^{(n)}{}_\theta(\beta, A, ..., \gamma_m) \times L(y / \beta_i^*, A_j^*, ..., \gamma_{l\,m}^*) \times d\beta \times dA \times .. \times d\gamma_m}, \qquad \textbf{(21)}$$

for $(\beta, A ..., \gamma_m) \in b_{i,j,...l}$,

where $\beta_i^*, A_j^*, \gamma_{k1}^*, \ldots, \gamma_{1m}^*$ are the midpoints in block $b_{i,j,k,..l}$, that is $\beta_i^* = \beta_1 + \frac{(2\times i-1)\times\Delta_1}{2}$, and similarly for the other parameters. The updating process for a two parameter space is shown in Figures 14 through 18.

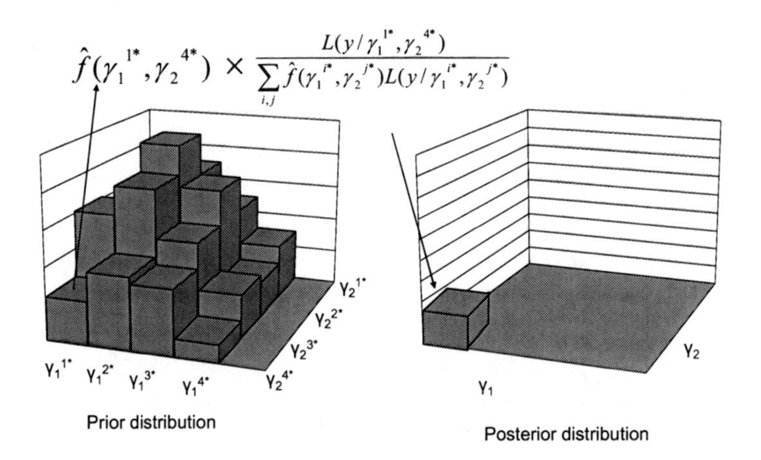

Figure 14: Prior distribution is updated using Bayes' rule: Part 1

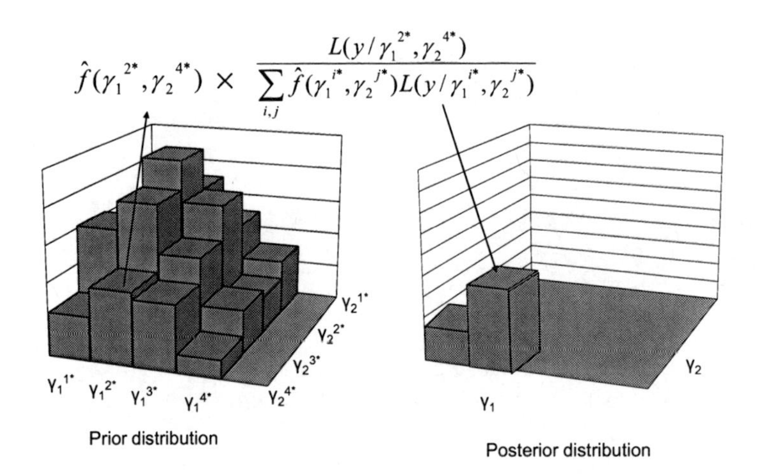

Figure 15: Prior distribution is updated using Bayes' rule: Part 2

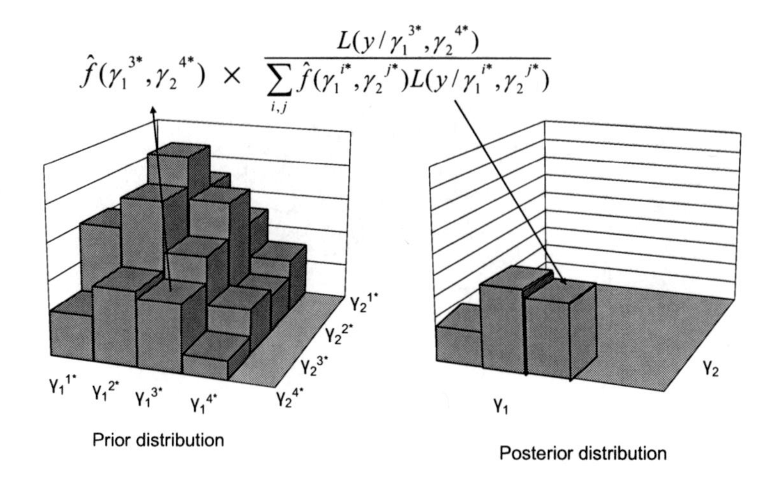

Figure 16: Prior distribution is updated using Bayes' rule: Part 3

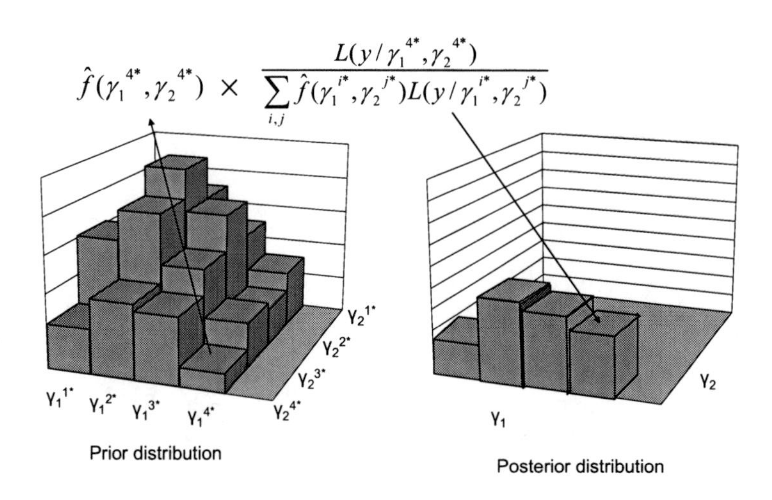

Figure 17: Prior distribution is updated using Bayes' rule: Part 4

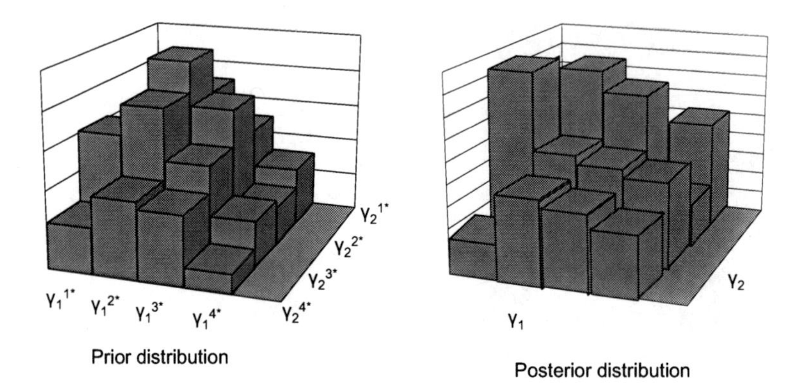

Figure 18: Prior distribution is updated using Bayes' rule: Part 5

In using this information to make a decision, we have a few options as described below:

(1) **(22)**

$$\hat{\beta} \approx \sum_{i,j,k,\ldots,r} \beta_i^* \times \hat{g}_{\beta,A,\gamma_1,\ldots,\gamma_m}(\beta^*, A^*, \gamma^*{}_1,\ldots,\gamma^*{}_m) \times V$$

$$s.d.(\beta) \approx \sqrt{\sum_{i,j,k,\ldots,r} (\beta_i^* - \hat{\beta})^2 \times \hat{g}_{\beta,A,\gamma_1,\ldots,\gamma_m}(\beta^*, A^*, \gamma^*{}_1,\ldots,\gamma^*{}_m) \times V}$$

(2) **(23)**

$$\hat{\beta} \approx \frac{\sum_{i=1}^{n} \beta^i L(y / \beta^i, A^i,\ldots,\gamma_m^i)}{\sum_{i=1}^{n} L(y / \beta^i, A^i,\ldots,\gamma_m^i)}$$

$$s.d.(\beta) \approx \sqrt{\frac{\sum_{i=1}^{n} (\beta^i - \hat{\beta})^2 L(y / \beta^i, A^i,\ldots,\gamma_m^i)}{\sum_{i=1}^{n} L(y / \beta^i, A^i,\ldots,\gamma_m^i)}}$$

The previous method can be used to estimate moments of other parameters as well. Formula 23 provides more accurate results since it is derived directly from the sample data which means it has less approximation.

So we can approximate the decision function using the above estimates:

$$C(x,\beta,A,\gamma_1,\ldots,\gamma_m) \approx C(x,\hat{\beta},\hat{A},\hat{\gamma}_1,\ldots,\hat{\gamma}_m)$$

(3) We also can use the modes of the posterior distribution as estimations for the parameters. We call them the maximum likelihood estimator (MLE) which is similar to the method introduced in section 9.3 for prior distribution. To find them we need not calculate the denominator in formula 21. Since the prior distribution is constant over each block and in this approximation the likelihood is also constant over each block the computation is easier. To be more accurate we can also use the average of likelihood function of the sample data in each block as the likelihood value of that block. This includes less approximation than the above method. After finding the block that gives us the highest value for the posterior distribution we find values for the parameters in that block that result in the highest value for the likelihood function. This sample is considered to be MLE of the posterior distribution. To see how this method works, please see the Dofasco case study.

If we want to obtain a better estimate for our decision function at the expense of a greater mathematical burden we have the following choices:

1) $$C(x) \approx \int_{\beta,A,\gamma_1,\ldots,\gamma_m} C(x,\beta,A,\gamma_1,\ldots,\gamma_m)\hat{g}_{\beta,A,\gamma_1,\ldots,\gamma_m}(\beta,A,\gamma_1,\ldots,\gamma_m)d\beta \times dA \times d\gamma_1 \times \ldots \times d\gamma_m \quad \textbf{(24)}$$

2) $$C(x) \approx \frac{\sum_{i=1}^{n} L(y / \beta^i, A^i, \gamma_1^i,\ldots,\gamma_m^i) \times C(x,\beta^i, A^i, \gamma_1^i,\ldots,\gamma_m^i)}{\sum_{i=1}^{n} L(y / \beta^i, A^i, \gamma_1^i,\ldots,\gamma_m^i)} \quad \textbf{(25)}$$

The second approximation is more accurate because it comes directly from the sample data.

11. MODEL CHECKING

Checking the model is crucial to statistical analysis. Bayesian prior to posterior inferences consider the whole structure of a probability model and can yield false inferences when the model is invalid. A good Bayesian analysis, therefore, should include at least some check of the adequacy of the fit of the model to the data and the plausibility of the model for the purposes for which it will be used (Gelman et al., 2004). The question in the model checking stage is: Do the model's deficiencies have a visible effect on the substantive inferences? In the next three sections we will address this legitimate concern.

11.1 Testing the Procedure in the Presence of Perfect Knowledge

In this experiment a graduate student with some advanced knowledge in statistics and probability was selected, and asked to choose values of parameters of a Weibull PHM with two covariates. Then he was asked to select some arbitrary values (within specified ranges) for the parameters of the Weibull PHM he had chosen. Afterward he was given 17 questions (questions similar to those designed to elicit expert knowledge in industry with some differences) and was asked to answer them based on the model he had selected. Once he had answered all the questions, the prior distribution for the parameters was calculated as described in Chapter 9. For each question an inequality which consists of the parameters of the model was developed. Therefore in this case we ended up with 17 inequalities. Using MATLAB software, a few thousand random samples of the parameters' values conforming to all 17 inequalities were taken. Then a histogram for each of the parameters was plotted. The results are shown in figure19.

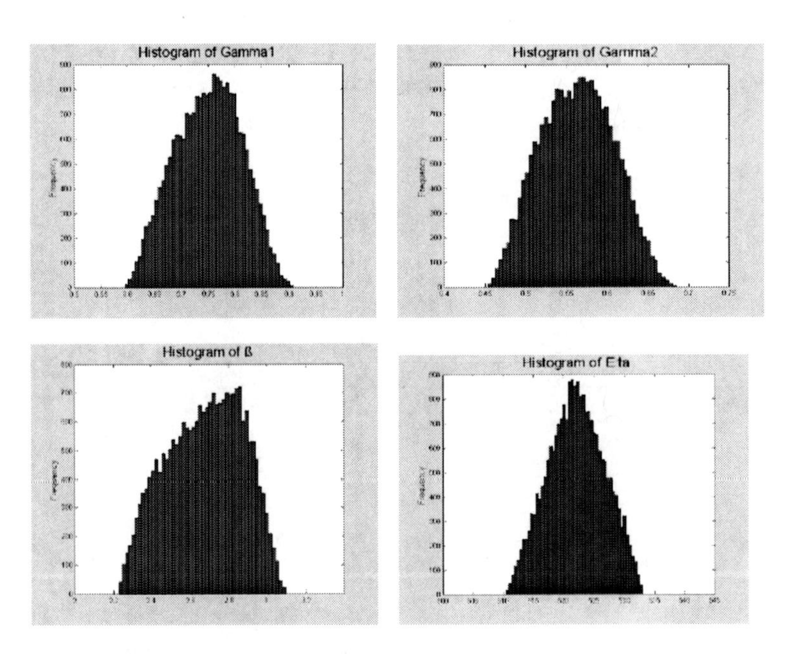

Figure 19: Histograms of ß, ŋ, γ_1, and γ_2

The results were compared with their real values (values selected by the subject) in Table 6. As can be seen, it shows that if knowledge about the parameters of a PHM exists our methodology is capable of capturing it.

Table 6

Comparison between the Results of Knowledge Elicitation and Real Values

Parameters	Sample mean	Sample s.d.	Real value
β	2.6765	0.1967	2.62
η	522.0238	4.6359	520
γ_1	0.7506	0.0614	0.72
γ_2	0.5635	0.0445	0.56

11.2 Testing the Procedure in the Presence of Partial Knowledge

After testing the questions and the prior distribution building process in the presence of perfect knowledge, it was decided to simulate the real world where no perfect knowledge is available. For this purpose, MATOR (machine simulator) software was developed. The software simulates a machine whose failure rate depends on its age and two time-varying covariates (pressure and temperature). The machine uses a PHM formula for calculation of the hazard rate. The parameters of the PHM can be introduced in the software by the programmer. On the software screen the values of both covariates and the component's age are depicted. One can play with the software and make maintenance decisions (replace preventively or run until the next inspection) based on the values of the covariates, component's age, cost of failure, and cost of preventive replacement. One can let the component fail and incur the failure cost (C_f) or replace the item before it fails and incur the cost of preventive replacement (C_p). In both cases the total cost and the average cost per unit of time of all previous histories is calculated and shown to the player. One of the screen windows is shown in Figure 20. A few volunteers were selected for this experiment. They were asked to do their best to minimize the average cost per unit of time. Then they were interviewed to elicit their experience about the failure mechanism of MATOR. In each run, all the condition monitoring data and events data were stored in a Microsoft Excel file so that other software packages could also be used to analyze the data. After bounding inequalities were created, MATLAB was run for two hours to find a few thousand samples that conform to all the inequalities. The data obtained from MATOR were input to EXAKT (Banjevic et al., 2001; Vlok et al., 2002), a condition-based maintenance software package based on a PHM that has been developed at the CBM Lab at the University of Toronto (see section 11.3.1). Then the results obtained from expert knowledge and the data processed by EXAKT were compared. Since the results obtained from the two sources matched adequately we consider the knowledge elicitation methodology to be effective. After this initial experiment we decided to test the performance of the methodology using a real industrial case.

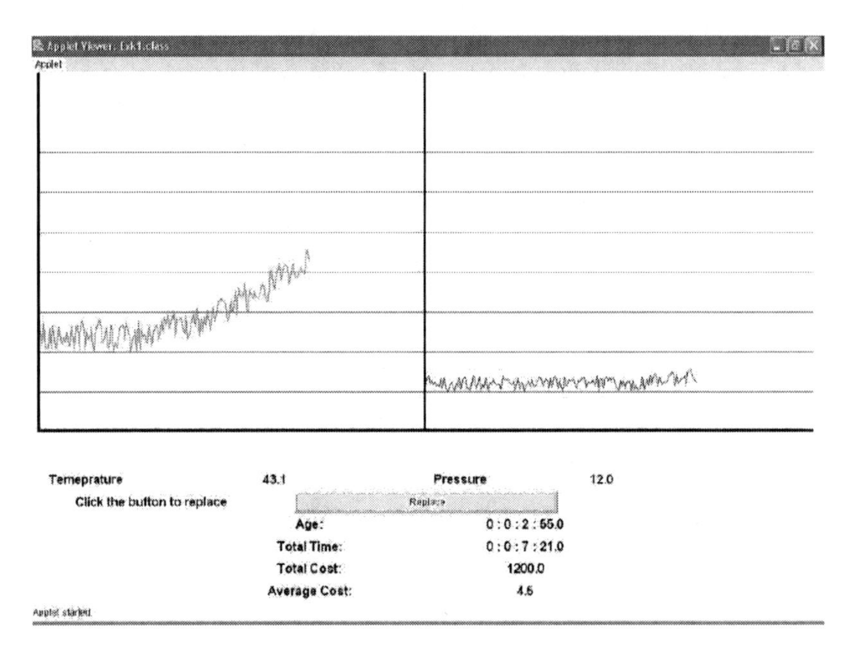

Figure 20: A screen window from MATOR

11.3 Testing the Model in Industry

The failure process of 3rd-stage piston rings (Figure 24) of Bellis and Morcom Compressors (Figure 25) at Dofasco Steel Manufacturing was chosen for analysis in 2004. Dofasco Steel is located in Hamilton, Ontario (Canada). This particular case was chosen because in 2002 a Condition-based maintenance model using EXAKT software had been developed successfully for this component. That model was based on 39 histories (26 failed, 7 censored, 6 that were operational at the time of the study) which was sufficient data to estimate PHM parameters with satisfactory precision. In other words, a reliable model existed which could enable us to compare results obtained by our methodology with the results obtained using the data set.

In fact, EXAKT was the main motivation behind this book. The development of this software package was started in 1995 with support from several companies including Dofasco. Since then it has been applied in many real industrial cases. A major problem that EXAKT faces is its need for lots of statistical data in a special format. The need for a methodology that can use expert knowledge and data together in estimation of the parameters of EXAKT resulted in defining this book. That is why it seems reasonable to know more about EXAKT before going into more detail regarding the case study undertaken at Dofasco.

11.3.1 What is EXAKT?

Use of PHM in reliability goes back to the 1980s (see section 3.1). The idea of using PHM was an outgrowth of a project done in the Canadian Department of National Defence (DND) (Anderson et al., 1982). To decide when to replace Pratt and Whitney engines on the Boeing 707, the engineers at DND measured some engine oil deposits, and the rates of change of those deposits were then compared to some standards. If the rate exceeded its limit, the engine was removed from service and sent to Pratt and Whitney for major overhaul. However it was noted that in some cases engines that had been removed were in fact in good condition and could serve on the aircraft for a further period. This created an interest in obtaining a good estimate of what the risk was of an engine transitioning to a failed state between two scheduled inspections (Jardine, 2002). If this risk was considered high then the engine should be removed; if it was seen as a low risk, it should be left on the aircraft until the next scheduled inspection. Within a few years following this initial study PHM was used in more applications (Jardine & Anderson, 1985; Jardine et al., 1989; Jardine et al., 1987).

The next step was to incorporate cost consideration in the model. In this way replacement decisions could be made based on the consequence of a failure compared to the consequence of preventive replacement. Obviously one tends to be more cautious when the cost of a failure is, say, 20 times higher than cost of preventive maintenance compared to the case where this cost ratio is, say, 2. The question is how these replacement decisions can be made to result in minimum cost in the long run. Makis and Jardine (1992) developed a decision policy to answer this question (Makis & Jardine, 1991a; Makis & Jardine, 1991b). Their decision policy is defined as a rule for replacement or leaving an item in operation until the next decision opportunity, depending on the results of the condition monitoring (Jardine et al., 1997). The criterion they used in their model was to choose an optimal policy that results in minimum average maintenance cost per unit of time. Those costs are due to the cost of preventive replacements and the cost of failure replacements. To calculate the hazard function they used a PHM with Weibull baseline hazard and time-dependent covariates as shown below.

$$h(t, Z(t)) = \frac{\beta}{\eta}\left(\frac{t}{\eta}\right)^{\beta-1} \exp(\gamma_1 Z_1(t) + ... + \gamma_m Z_m(t))$$

where:

$\frac{\beta}{\eta}\left(\frac{t}{\eta}\right)^{\beta-1}, \gamma, Z(t)$ are defined in equation 1.

Incorporation of a cost model for optimal decision-making requires predicting the future values of the hazard function. This in turn depends on the stochastic behavior of the covariates. In the model developed by Makis and Jardine (1992) it was assumed that covariates follow a non-homogeneous Markov stochastic process. The transition probabilities of the Markov chain are calculated based on the data (see Makis & Jardine, 1991a). It is also assumed that the item deteriorates with time, which means that failure risk increases with time and covariates. Makis and Jardine (1992) showed that the optimal replacement policy can be found in a class of stopping times. If we consider C as the preventive replacement cost, C+K as the failure replacement cost, and T_d, d>0, as the stopping time of the form:

$$T_d = \inf\{t \geq 0 : h(t, Z(t)) \geq d / K\}$$

If we define T as the random variable representing time to failure, and

$Q(d) = P(T_d \geq T)$, $W(d) = E(\min\{T_d, T\}$, then $Q(d)$ is the probability that failure replacement will occur, using a strategy with hazard risk level *d/K*. $W(d)$ is the expected time until replacement either due to a preventive replacement or a failure replacement (Jardine et al., 1997). As a result, the expected average cost per unit time is: $\Phi(d)=\frac{C+KQ(d)}{W(d)}$

The optimum hazard level is called d^* and it is defined by:

$$\Phi(d^*)=\min_{d>0}\Phi(d)=d^*$$

This means that if one replaces the component when its hazard level reaches d^*/K, in the long run, the average cost per unit of time will be minimized.

Q(d), *W(d)*, and $\Phi(d)$ are calculated based on a recursive procedure developed by Makis and Jardine (1991a). After calculating $\Phi(d)$, it would be easy to find the optimal value for d^* using the following recursive formula: $d_n=\Phi(d_{n-1}),\ n=1,2,...,\ d_0>0$

This is equivalent to $d^*=\lim_{n\to\infty}d_n$ (Makis & Jardine, 1992). This iterative process usually converges after a few steps (Jardine et al., 1997).

Once d^* is calculated, to arrive at the minimum average cost per unit time, one has to replace the item at the first moment t, when

$$\frac{\beta}{\eta}\left(\frac{t}{\eta}\right)^{\beta-1}\exp(\gamma_1 Z_1(t)+...+\gamma_m Z_m(t))\geq\frac{d^*}{K}.$$

This is equivalent to replacing the item at the first moment t for which

$$\gamma_1 Z_1(t)+...+\gamma_m Z_m(t)\geq\delta^*-(\beta-1)\ln t,$$

$$\delta^*=\ln\frac{\eta^\beta d^*}{\beta K}$$

EXAKT uses a graphical demonstration to show decisions regarding replacement (Figure 21).

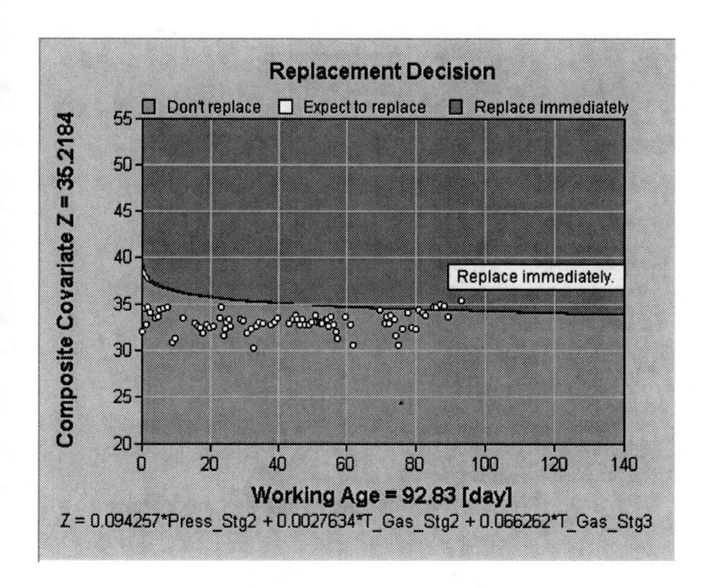

Figure 21: Decision-making graph of EXAKT for a sample component

Measurements from each inspection are mapped into a graph similar to the one shown above, if a point falls in the green zone, it means we should continue using the item until the next inspection, if it falls in the red zone it means we have to replace the item immediately. There is usually a small yellow area that implies that the component should be replaced before the next inspection.

11.3.1.1 Structure of EXAKT

So far we have explained the mathematical theory behind EXAKT. In reality there are many additional issues that have to be resolved before one can apply any theoretical model to real industrial cases. In this regard EXAKT is not an exception. Its main problem is the lack of enough reliable data in proper format. For example, EXAKT needs to know when a component was installed, what maintenance actions have been done during the component's life, the values of condition indicators (covariates) during its life, when the component was removed, and for what reason. Many companies use computerized maintenance management systems (CMMS) to collect data, schedule maintenance activities, measure the number of spare parts in the plant's inventory, and collect and keep the condition monitoring data and histories of maintenance activities. However the data gathered by CMMS has to be integrated, analyzed, and cleaned before using them in EXAKT. This process can be a very time consuming. To make the process easier, EXAKT has some data conversion and pre-processing tools. These tools help the user transform the data obtained from CMMS or other software packages to the format required by EXAKT (Figure 22)

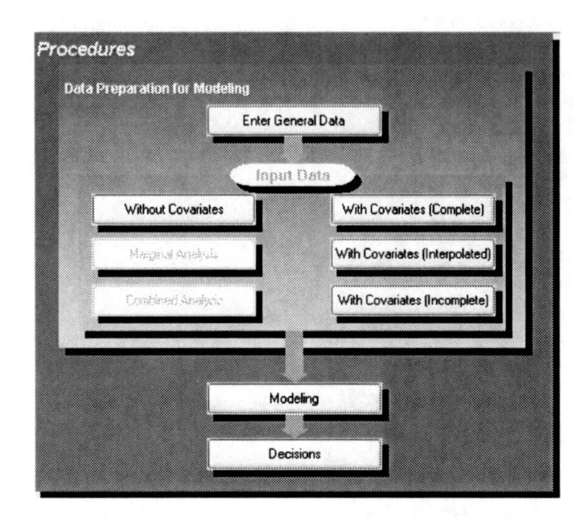

Figure 22: EXAKT's main components at glance

When cleaned data are fed to EXAKT, the software calculates the parameters of PHM and transition probabilities using ML. To find the best replacement policy, EXAKT needs cost information as well. It needs to know the cost of preventive replacement and failure replacement. Then using the methodology described earlier, it calculates the optimum policy.

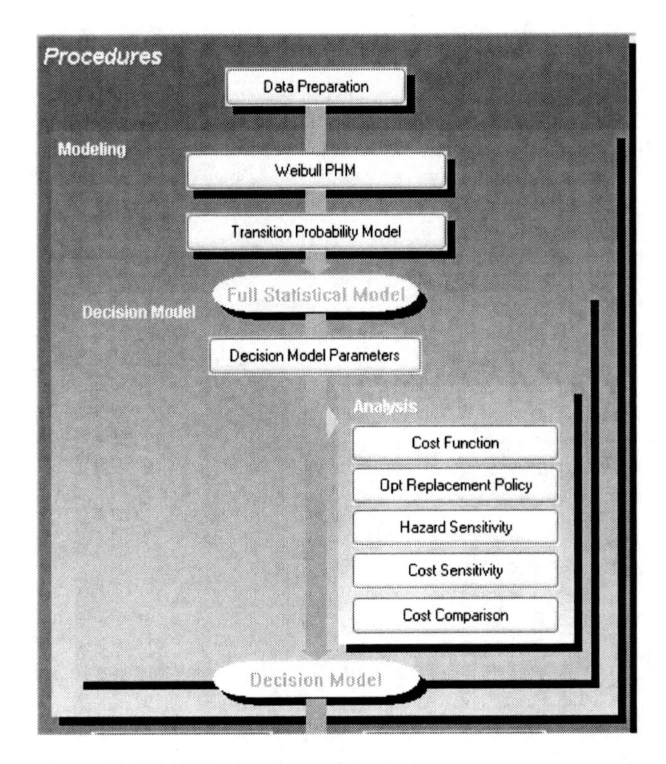

Figure 23: EXAKT's decision-making features

To fund development of the software and also to create a support from industry in getting real data and applied experiences regarding condition-based maintenance practices, the Condition-based Maintenance Laboratory was created at the University of Toronto in December 1995. Financial support for the first three years was provided by the Natural Sciences and Engineering Research Council of Canada, Material and Manufacturing Ontario, and the following six companies:

1. Alcoa
2. Barrick Gold
3. Dofasco Steel
4. Molson Breweries
5. Syncrude Canada
6. Wear Check Canada

Over time new companies joined the consortium while others terminated their support. Currently a group of 11 companies actively support the laboratory. Now that we know more about EXAKT and its decision-making strategy we return to the case study.

11.3.2 What is the Problem?

The failure process of 3rd-stage piston rings (figure 24) of Bellis and Morcom Compressors (figure 25) is considered in this case study. These compressors are reciprocating type compressors and are located at the nitrogen plant in Dofasco. Reciprocating compressors are widely used in industry; therefore the results and insights obtained from this study are likely to be useful to many industries (Zyl, 2002). Based on the definition, a piston ring would be classified as failed if the radial thickness is less than the minimum specified by the vendor (0.268in). At present, Dofasco has 28 condition indicators for each compressor. Inspections are carried out in intervals between 12 and 24 hours. Not all the indicators are independent; some are functions of other condition indicators. These condition indicators are used to predict failures of a few components such as intercoolers, valves, heat exchangers, and the piston rings.

Figure 24: 3rd-stage piston rings, old (left) and new (right)

Figure 25: Bellis and Morcom compressors

The expert selected for this case study was responsible for the maintenance of the Bellis and Morcom compressors in the nitrogen plant. He does not have statistical knowledge and was not part of the project in which EXAKT was used to estimate the parameters of the PHM for those compressors in 2002. Therefore he had no clear knowledge regarding the statistical results of that case study. The knowledge elicitation process explained in chapter 8 was used to extract his knowledge regarding the failure mechanism of the 3rd-stage piston rings of the compressors. He believes that the following covariates are the key indicators of the health condition of the Bellis and Morcom compressors (Figures 26–28):

1. 2nd-stage discharge gas pressure (psi),
2. 2nd-stage discharge gas temperature (F),
3. 3rd-stage discharge gas temperature (F), and
4. Age of the piston (days).

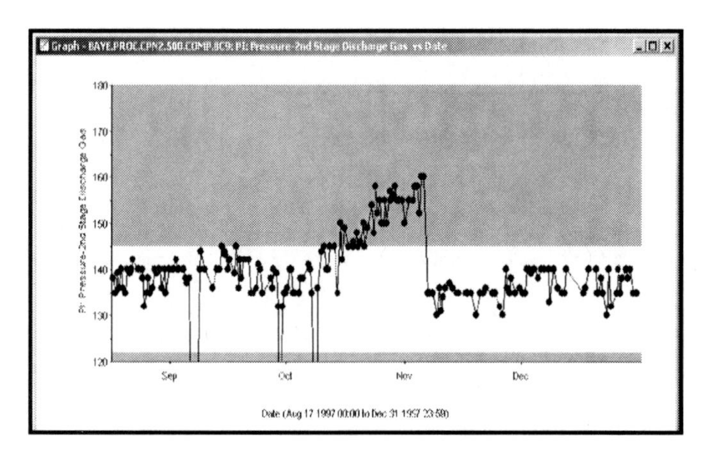

Figure 26: Monitoring pressure of the 2nd-stage discharge gas

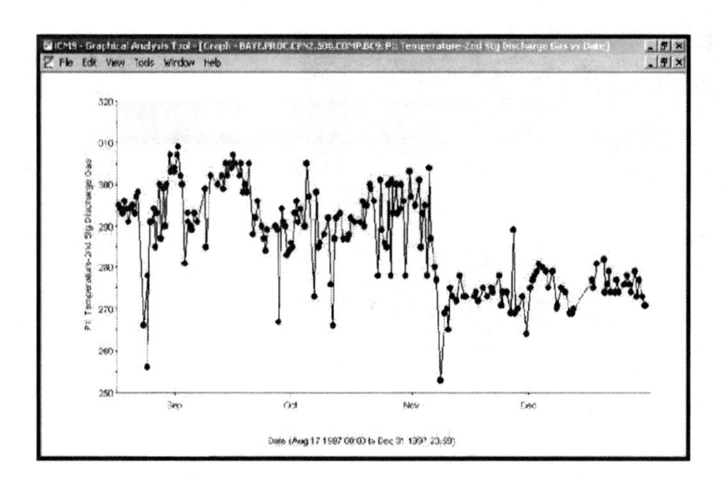

Figure 27: Monitoring temperature of the 2nd-stage discharge gas

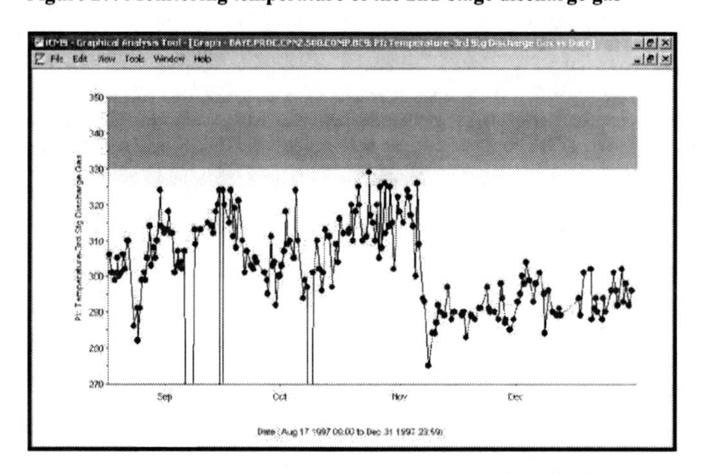

Figure 28: Monitoring temperature of the 3rd-stage discharge gas

The methodology which was tested using MATOR was modified slightly to take this real case into account. Knowledge elicitation began with some general questions regarding the failure process of the compressors, and then focused on 3rd-stage piston rings. Some of these questions along with the expert's answers are shown below. Note that the format of the case comparison questions in this case study is slightly different from what was shown in section 8.3. The format is subject to change depending on the expert, the failure process, and other factors.

11.3.3 Primary Questions

What is the average life of the piston ring?

Between 3 and 5 months

What are the longest and shortest life spans for the piston ring?

Shortest lifetime: 2.5 months
Longest lifetime: 5.5-6 months

What are the key indicators of the failure of the 3rd-stage piston rings?

1. 2nd-stage discharge gas pressure
2. 3rd-stage discharge gas temperature
3. 2nd-stage discharge gas temperature

Which of them is more important?

First: 2nd-stage discharge gas pressure
Second: 3rd-stage discharge gas temperature

What are the meanings of the different control limits of condition indicators?

Warning limit:
Let's watch this component; something is changing with the equipment
Alarm limit:
Get ready for intervention; look at PM work order and try to take action in the next 3-4 days
Critical level:
Immediate action, today or tomorrow

What are control limits of the key condition indicators of the 3rd-stage piston ring?

1. 2nd-stage discharge gas pressure
   Normal: 126-142 psi
   Warning: 142-145 psi
   Alarm: N/A
   Critical: 145 psi

2. 3rd-stage discharge gas temperature
   Normal: 250-325 F
   Warning: 325-330 F
   Alarm: N/A
   Critical: 330 F

3. 2nd-stage discharge gas temperature
   Normal: 200-310 F
   Warning: above 310 F
   Alarm: N/A
   Critical: N/A

As explained in section 8.3, primary questions help us design sensible cases for the case comparisons and case analyses section of the knowledge elicitation. Based on the methodology described in section 8.3.1 we developed some cases for analysis and comparison by the expert. To see a complete list of case comparison questions see Appendix A.

11.3.4 Case Comparison Questions

11.3.4.1 Type 1 Question

Please compare cases A and B in each table in terms of the probability of having a failure (risk of failure). Please also assume that other components such as valves and intercooler system are working properly, and we only consider failures of the 2nd-stage piston ring. Please try to follow terminology shown below when comparing the two cases:

- Case A (B) has much higher risk of failure compared to Case B (A).
- Case A (B) has higher risk of failure compared to Case B (A).
- Case A (B) has slightly higher risk of failure compared to Case B (A).
- Case A (B) has almost the same risk of failure compared to Case B (A).
- These cases do not make any sense based on my experience.
- Or any other explanation that you think is appropriate which I have not mentioned here.

Case 1:

	Case A	**Case B**
2nd-stage discharge gas temperature (F)	320	310
3rd-stage discharge gas temperature (F)	325	330
2nd-stage discharge gas pressure (psi)	140	140
Age (months)	4	4

Answer: Case A and Case B have the same risk.

Case 2:

	Case A	**Case B**
2nd-stage discharge gas temperature (F)	300	310
3rd-stage discharge gas temperature (F)	325	330
2nd-stage discharge gas pressure (psi)	150	140
Age (months)	4	4

Answer: Case A has slightly higher risk of failure compared to Case B.

11.3.4.2 Type 2 Question

Please compare cases A and B in each table in terms of the probability of having a failure (risk of failure). Also assume that other components such as valves and intercooler system are working properly and we only consider failures of the 2nd-stage piston ring. What makes this case comparison different from previous case comparisons is that it is obvious that Case B is at higher risk of failure (since the values of the condition indicators in Case B are more than or equal to those of Case A). The question is how many times Case B is at greater risk of failure compared to Case A. For example you can say that Case B is between 2-4 times riskier compared to A. Use the upper and lower limits which make you 90% confident that the right value would be between them. For example when you say Case B is between 2 and 4 times riskier than Case A, this should also imply that you are 90% sure of your answer. If you are not sure you can expand the interval to a point that makes you confident. For instance in this case if you think the interval between 2 and 4 does not make you 90% confident, you can use 1.5 and 4.5 as lower and upper limits.

Case 1:

	Case A	**Case B**
2nd-stage discharge gas temperature (F)	300	310
3rd-stage discharge gas temperature (F)	325	330
2nd-stage discharge gas pressure (psi)	140	140
Age (months)	4	4

Answer: B is between 1.1 and 1.2 times riskier than A.

Case 2:

	Case A	**Case B**
2nd-stage discharge gas temperature (F)	300	310
3rd-stage discharge gas temperature (F)	320	330
2nd-stage discharge gas pressure (psi)	135	140
Age (months)	4	4

Answer: B is between 2 and 3 times riskier than A.

The expert was asked to compare around 45 cases similar to those shown above. His answers to the questions created around 60 bounding inequalities. Based on the inequalities the upper and lower limits for the parameters were obtained. Then 50,000 samples were simulated which resulted in estimates of the parameters. These results are shown and compared with the results obtained from the data set using maximum likelihood in Table 7.

Table 7

PH Parameters Obtained Based on Expert Knowledge and Those Calculated Using Statistical Data

	Lower limit	Upper limit	Estimated value from expert knowledge	Standard deviation	MLE of statistical data	Standard deviation
β	1.5000	1.7998	1.559286	0.0520	1.8620	0.3192
A	36.3993	44.8467	41.1004	1.4098	40.4008	7.61
γ_1	0.0814	0.1380	0.1214	0.0064	0.0874	0.03704
γ_2	0.0047	0.0090	0.0070	0.0012	0.0032	0.01615
γ_3	0.034511	0.064152	0.0540	0.0056	0.0648	0.01881

where

γ_1 is the coefficient of the second stage discharge gas pressure

γ_2 is the coefficient of the second stage discharge gas temperature

γ_3 is the coefficient of the third stage discharge gas temperature

As can be seen, the difference between values of parameters obtained from expert knowledge and those obtained using statistical data are relatively small based on the standard deviation of the parameters estimated using the method of maximum likelihood (the difference between estimates based on the two methods are usually less than one standard deviation of the estimates).

11.3.5 Analyzing the Results of the Experiments

In PHM with time-dependent covariates when a covariate has a strong correlation with time, the value obtained for β using maximum likelihood (ML) would be higher than its actual value. For example, assume an extreme case when one of the covariates is ln(t):

$$h(t;Z(t)) = h(t;Z(t)) = t^{\beta-1}\exp(\gamma_1 \ln(t) + ... + \gamma_m Z_m(t) - A)$$

$$= t^{\beta-1} t^{\gamma_1} \exp(\gamma_2 Z_2(t) + ... + \gamma_m Z_m(t) - A)$$

$$= t^{\gamma_1+\beta-1} \exp(\gamma_2 Z_2(t) + ... + \gamma_m Z_m(t) - A)$$

ML would increase β to $\beta+\gamma_1$ to incorporate the effect of this covariate. Another issue with ML is the collinearity problem which happens when covariates are highly correlated. Sometimes it gives a negative value to the coefficient of the covariate that has an increasing effect on the hazard. The issues mentioned above can prevent us from seeing very similar results obtained from the two methods when such cases occur. Since we do not know when the above situations (such as collinearity, age, or correlated covariate) appear, the question is how can we conclude that our methodology gives satisfactory results. An answer is that when the hazard rates obtained from the two methods give close outputs for the same data, we can conclude that the two methods are equivalent.

To test the model based on the suggested evaluation method, the value of the hazard rate for 60 randomly selected cases from the data set were calculated. Two examples are presented here:

(1)

2nd-stage discharge gas pressure	142 psi
2nd-stage discharge gas temperature	310 F
3rd-stage discharge gas temperature	330 F
Age of the piston	140 days

$h_E = h_{E+D0} = 0.321383$

$h_{D39} = 0.255192$

$$\frac{h_{E+D0}}{h_{D39}} = 1.2594.$$

h_{E+D0} is the hazard rate based on the expert knowledge combined with zero data history (it is also equivalent to the hazard rate based on the expert knowledge) and h_{D39} is the hazard rate based on MLE of all 39 data histories. This measure (here for example $\frac{h_{E+D0}}{h_{D39}}$) is called *measure of accuracy* in this book.

2nd-stage discharge gas pressure	126.9 psi
2nd-stage discharge gas temperature	280.7 F
3rd-stage discharge gas temperature	290.9 F
Age of the piston	100 days

(2)

$h_{E+D0} = 0.004211$

$h_{D39} = 0.003691$

$$\frac{h_{E+D0}}{h_{D39}} = 1.1408.$$

Similar outcomes were obtained for other selected records which resulted in the following median for $\frac{h_{E+Dj}}{h_{D39}}$ measure of accuracy where Dj represents j data histories and E represents expert knowledge.

$$\text{Median}\left(\frac{h_{E+D0}}{h_{D39}}\right)=1.8054$$

To show that the hazard calculated based on the estimated parameters follows the hazard calculated using the parameters based on 39 data histories, we measured the correlation of the two for all 60 randomly selected cases. High correlation (close to one) between two variables means that they usually move in the same direction. For this case we have:

$$\text{Correlation}\ (h_{E+D0}, h_{D39}) = 0.9883$$

Therefore, we consider that our methodology is capable of capturing expert knowledge. Obviously, due to the nature of expert knowledge and the inability to capture all relevant information, we cannot expect to obtain a very accurate model. The model based on expert knowledge should be used as a basic starting model that will be improved and updated as real data arrives. However, the significance of the above values of median of $\frac{h_{E+D0}}{h_{D39}}$ and correlation between h_E (or equivalently h_{E+D0}) and h_{D39} is an important issue which needs attention. To deal with this matter we picked the first ten of the 39 histories in the statistical data set. We found the MLE of the parameters based on those ten histories and achieved following numbers:

$$\beta = 1.2420$$
$$A = 45.8648$$
$$\gamma_1 = 0.1418$$
$$\gamma_2 = 0.0000$$
$$\gamma_3 = 0.0725$$

We calculated $\frac{h_{D10}}{h_{D39}}$ for all 60 randomly selected cases. h_{D10} is the hazard rate based on the MLE of 10 data histories.

This resulted in:

$$\text{Median}\left(\frac{h_{D10}}{h_{D39}}\right) = 1.9176$$

$$\text{Correlation}\ (h_{D10}, h_{D39}) = 0.9854$$

Although no generalization can be made based on only one case, the above result shows that parameters can be estimated more accurately based on expert knowledge compared to estimation achieved by 10 data histories in this particular case study.

If we combine the expert knowledge with these ten data histories using formula 23 in Chapter 10 and the procedure described in the next section, the following results are obtained:

$\beta = 1.5562$
$A = 41.0780$
$\gamma_1 = 0.1217$
$\gamma_2 = 0.0069$
$\gamma_3 = 0.0539$

$$\text{Median}\left(\frac{h_{E+D10}}{h_{D39}}\right) = 1.8573$$

$$\text{Correlation}\ (h_{E+D10}, h_{D39}) = 0.9881$$

Where h_{E+D10} is the hazard rate based on the expert knowledge and ten data histories combined using Bayes' rule.

The accuracy of the parameters based on ten data histories has been improved due to a slightly lower median for our measure of accuracy and a higher correlation between h_{D39} and h_{E+D10}. Still, one might argue that a hazard ratio equal to 1.8054 is very high, that it is equivalent to 84% error, which is not acceptable in many practical cases. However to be able to judge the quality of such an estimate we should look at how these estimates affect the decision-making process in reality. As mentioned in section 11.3.1, we use PHM to measure the hazard, and then compare its value to a threshold value. If it is higher than the threshold value we decide to replace the component otherwise we let the equipment run until the next inspection. The threshold value itself is obtained through some mathematical optimization. This value depends on the cost of preventive replacement, the cost of a failure replacement, the values of the parameters of the PHM which describe the relationship between conditions of the machine and its hazard rate, and the behavior of the condition indicators (covariates) of a system. In practice, hazard is usually very low during most of the life of a component. It increases as the component deteriorates. At some point the deterioration process speeds up and causes the hazard to increase with a higher slope. Most condition-based maintenance is used during this last stage, trying to predict failure and prepare for it before it happens. Behavior of the hazard during the life of a piston ring in this study is shown below:

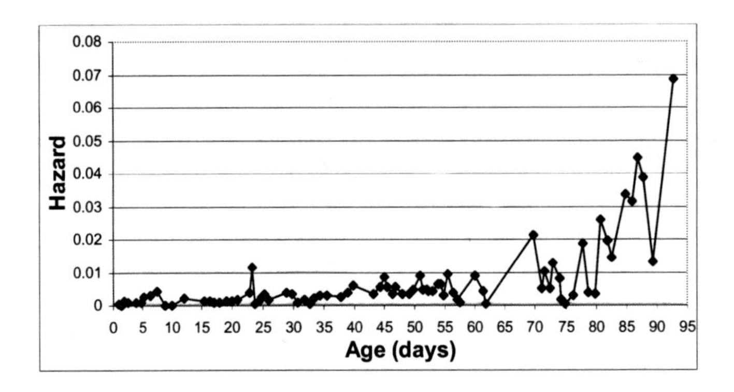

Figure 29: Behavior of the hazard during a life of a 3rd-stage piston ring

This hazard is calculated using the parameters estimated based on all 39 data histories which are our best estimate for the parameters. As can be seen, it seems that the deterioration process has been boosted around age 70. However there are lots of ups and downs in the behavior of the hazard. In practice condition monitoring specialists do not decide based on only one value, they use some trending and aggregation methods to make sure that the current increase is due to an incoming failure not just a spike in the indicators. It seems as though they smooth the data to see the trends. A similar effect can be obtained using a moving average of the data. If we use a moving average with a period of three, the above graph becomes as presented in Figure 30:

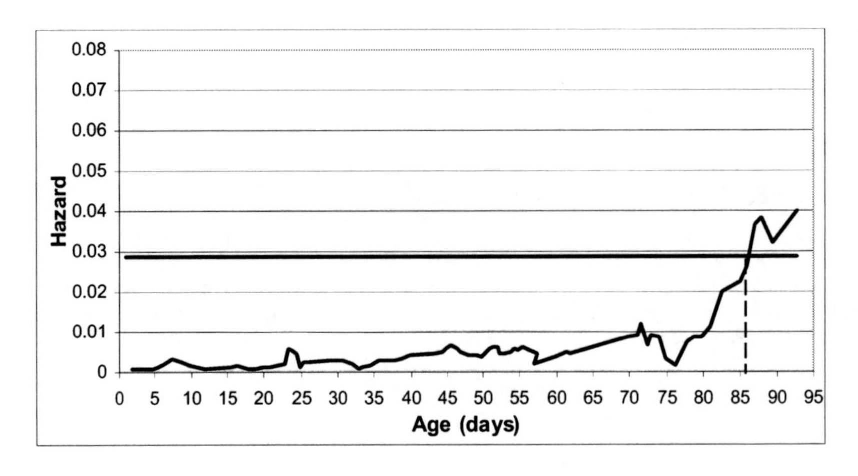

Figure 30: The component should be replaced when it reaches the optimum hazard level

The horizontal line at a hazard level of 0.0288 is the optimum hazard for replacing the component. It suggests that the item has to be replaced at age 86. If to the above graph we add the hazard calculated (moving averaged of course) using the parameters estimated based on the expert's knowledge we would have (the dotted lines in all figures below show the hazard estimated based on the expert's knowledge):

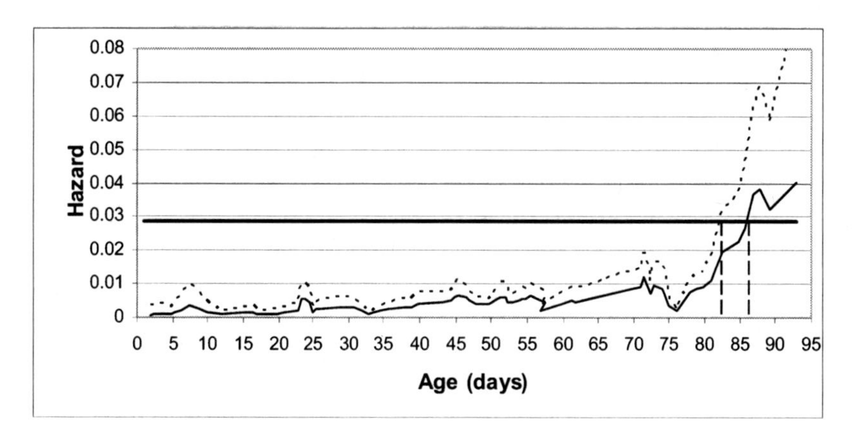

Figure 31: Hazard calculated based on the expert's knowledge and hazard based on the data results in different replacement times (1)

The optimum hazard level now suggests that the item should be replaced at age 82.5. This means that if we use the parameters based only on the expert's knowledge, we would change the component 3.5 days earlier. Similar results obtained during the life of other components some of which are shown in the next few graphs:

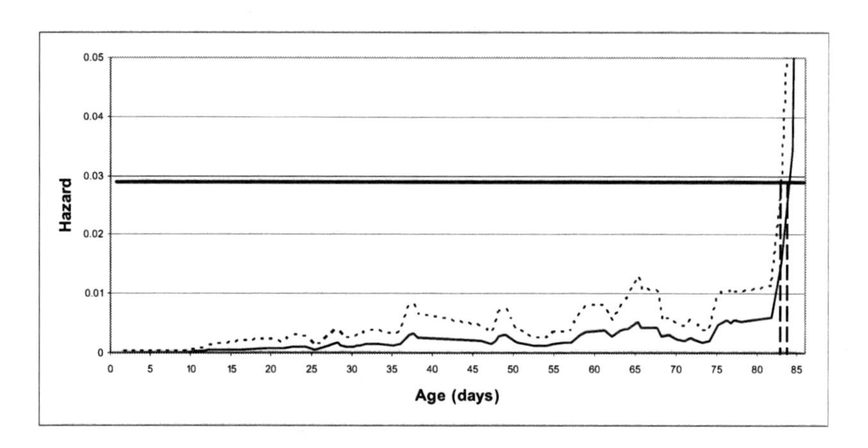

Figure 32: Hazard calculated based on the expert's knowledge and hazard based on the data results in different replacement times (2)

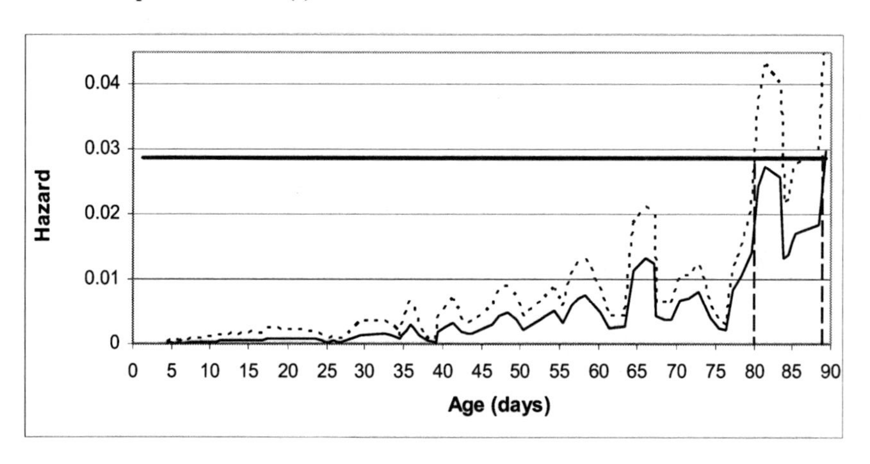

Figure 33: Hazard calculated based on the expert's knowledge and hazard based on the data results in different replacement times (3)

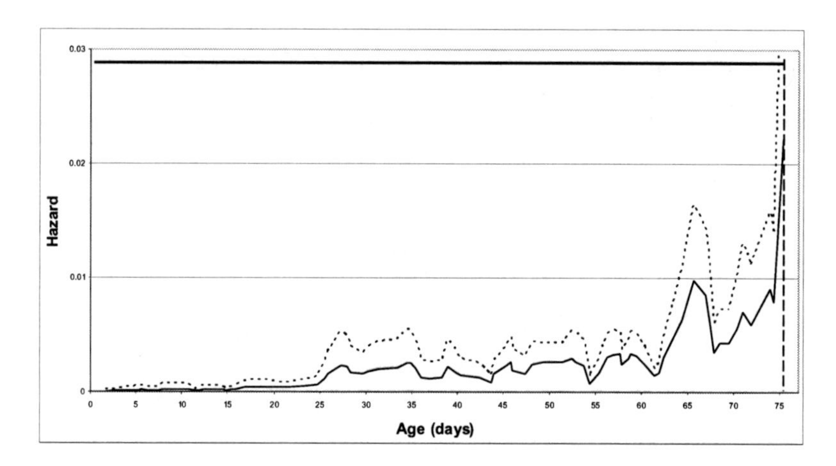

Figure 34: Hazard calculated based on the expert's knowledge and hazard based on the data results in different replacement times (4)

In Figure 32 the difference between replacement times is about 1.5 days, while in Figure 33 the difference is ten days. We should note that overestimating the hazard is not as bad as underestimating it. It is a more cautious and conservative way of measuring the hazard. For instance in Figure 34 it prevents a failure which could not be prevented using the hazard estimated based on the 39 data histories. By observing other histories it seems that making replacement decisions using the parameters estimated based on expert knowledge, the useful life of the components will be reduced very roughly by 3-6 %. Even if we ignore the savings due to being more conservative as shown in Figure 34, this reduction in average life will result in a maximum increase of 3.1%-6.4% in average cost per unit time. However this additional cost should be compared to the optimum average cost due to implementing software packages such as EXAKT. The result of the case study done at Dofasco in 2002 shows that if EXAKT had been applied, the average cost per unit time would have been reduced by 35% compared to the company's policy at the time. This leaves us with a saving of between 28.6% and 31.9%.

In this case study, the correlation between hazards calculated based on parameters obtained using expert knowledge and those obtained using data is very high which means that both hazards show similar behavior as can be seen in the preceding figures. Knowing that our methodology transfers the knowledge of the expert to the prior distribution we might want to find the source of the overestimation of the hazard. The range of the hazard is determined by the inequalities described in formula 9 in Chapter 8. So if we make extra effort when estimating upper and lower bounds for the hazard (P_u and P_l) it might reduce the overestimation. So far we have used the method of direct estimation of P_u and P_l. Some researchers (Garthwaite et al., 2005) suggest using observable or potentially observable quantities in knowledge elicitations. Based on this advice we thought that it might be worth trying to ask the expert to give us his best estimate for the remaining useful life (RUL) instead of the probability of having failure in an interval. A few cases were designed and shown to the expert. He was asked to give his best estimate for both the probability of having a failure in the next 24 hours and for the remaining useful life of the component under specific conditions. One of those questions with its instruction follows. (For a full list of these questions see appendix A.)

"Consider failure of 3rd-stage piston rings. Please assume that other components such as valves, intercooler system, etc are working properly and only consider failures of the 2nd-stage piston ring when the component has not yet failed. For each case please give your best estimate of probability of having a failure during the next 24 hours and also the average of remaining time to failure. You can provide intervals for the required answers, for example you can say: "I am 90 % confident that the item will fail in 2-5 days."

Cases are described below:

	Sample Case
2nd-stage discharge gas temperature (F)	310
3rd-stage discharge gas temperature (F)	330
2nd-stage discharge gas pressure (psi)	145
Age (months)	4.5

Probability of having a failure within the next 24 hours: 70

Mean time to failure from now (how long does it take in average for the component to fail from now): 2-3 days

Based on the first statement the expert indicated that the component will fail in the next 24 hours with a probability of 70%. At the same time he indicates that the component will fail in between 2 and 3 days or in an average of 2.5 days. If we approximately calculate the value of the hazard based on the two methods, the first statement estimates the hazard for the given conditions to be 1.2 while the second statement estimates it to be 0.4. As shown in this sample case, the direct estimation of the probability of failure estimate of the hazard is three times higher than using RUL. These results are consistent with the result indicated in (Seaver & Stillwell, 1983) which says that when the consequence of the failure is high the expert overestimates the probability of the event. After this observation we modified our model based on the expert's answers for remaining useful life. We obtained the following values for the parameters.

$\beta = 1.5568$

$A = 41.5107$

$\gamma_1 = 0.1215$

$\gamma_2 = 0.0070$

$\gamma_3 = 0.0539$

These parameters result in:

$$\text{Median}\left(\frac{h_E}{h_{D39}}\right) = 1.1842$$

$$\text{Correlation}\,(h_E, h_{D39}) = 0.9881$$

Now let us see how these new estimates would change Figures 31–34.

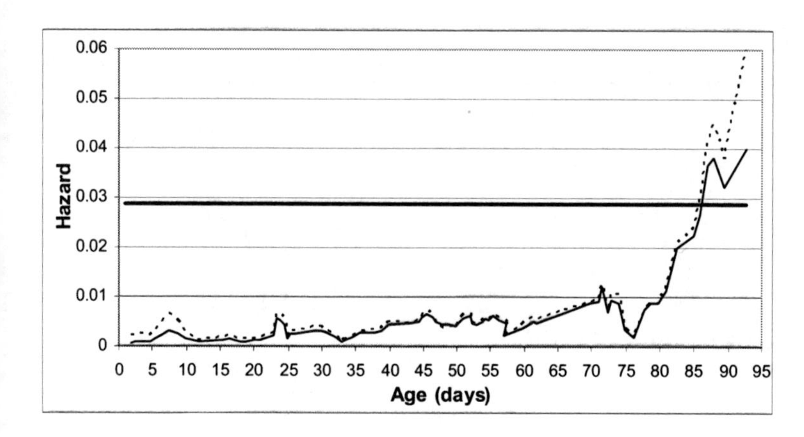

Figure 35: Figure 31 modified based on the new results

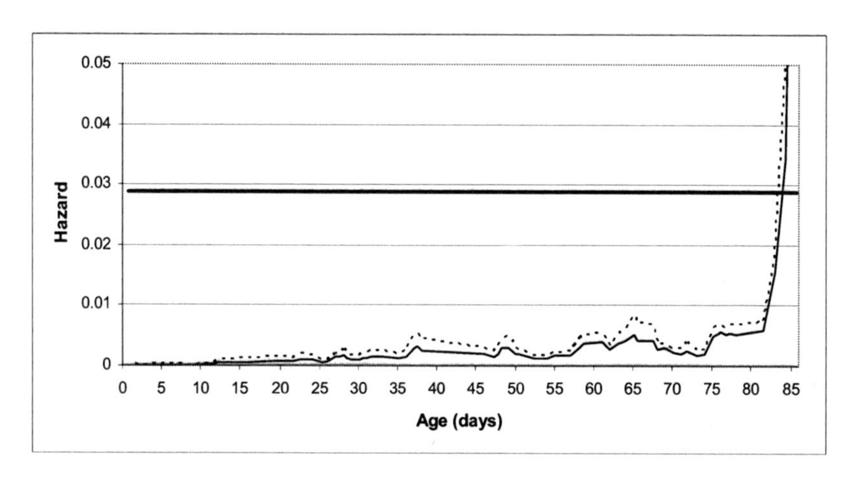

Figure 36: Figure 32 modified based on the new results

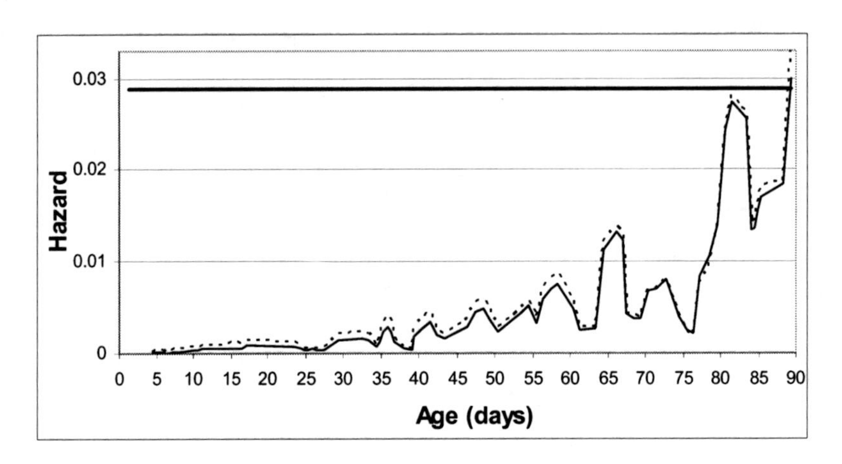

Figure 37: Figure 33 modified based on the new results

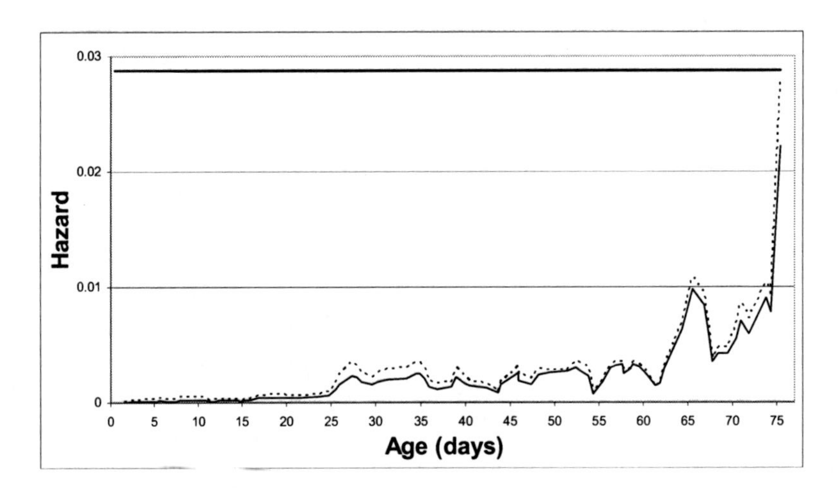

Figure 38: Figure 34 modified based on the new results

As shown in Figures 35 through 38, the difference between the recommended replacement times is very small (on average less than a day).

Note: The results obtained in this case study generally are more accurate than we expected. This may be due to the following reasons:

1. The statistical data is noisy; therefore, the maximum likelihood estimators (MLEs) based on these data do not converge fast enough and create wide confidence intervals. This also can be seen by looking at the MLEs based on all 39 histories and MLEs based on the first ten histories.
2. The expert has profound experience in his job; therefore, he has been able to provide high precision information. Highly experienced experts usually pay a lot of attention to the dynamics of the system and therefore are able to recognize the patterns and ignore noises. In contrast, statistical models can only work with data.
3. The knowledge elicitation process developed in this research reduces variability. For example if we obtain the same inequality twice based on the expert's answer and each of them is based on 90% confidence, this means that inequality will be active 99% ($1\text{-}0.1^2$) of sampling time which reduces the variability. It might be argued that we should not include such inequality twice in our model; on the other hand, one might say that obtaining the same inequality twice is good evidence that the inequality has more validity than inequality obtained only once. We have not been able to rule out either of these arguments completely. In early experiments (Zuashkiani et al., 2006) we used each inequality only once (in most cases), later on based on the second logic we decided to include all inequalities no matter how many times they show up. The results based on the latter approach have tighter confidence intervals as one might expect but the median value of measure of accuracy did not change considerably.

11.3.6 Updating the Prior Distribution to Posterior Distribution

To test the updating procedure described in Chapter 10, a computer program was developed to obtain the likelihood function of the data as a function of the PHM parameters. Since there are 4161 inspections and 26 failures, the likelihood function contains more than 4200 components. We used both methods of mean and MLE of posterior distribution described in Chapter 10 to calculate updated estimates of the parameters and both methods showed similar results. To find the MLE of the prior distributions, based on 50,000 samples generated in the previous section, upper and lower limits for each parameter were obtained. Each parameter was divided into ten equal parts resulting in 100,000 (10^5) blocks. Block $b_{0,2,6,2,8}$ contained 137 sample data which was the highest among all the blocks (here we consider the results of the first attempt for which upper and lower bounds for the hazard was extracted directly by asking the expert for the probability of having failure in a time period). The average value of data in $b_{0,2,6,2,8}$ for each parameter is:

$$\beta = 1.5162$$
$$A = 41.7317$$
$$\gamma_1 = 0.1177$$
$$\gamma_2 = 0.0057$$
$$\gamma_3 = 0.0596$$

Then each sample was multiplied by the value of the likelihood function for that sample and the results were stored in its corresponding block. This time $b_{1,2,6,2,8}$ contained the highest value and the sample data within $b_{1,2,6,2,8}$ which provided the maximum value for likelihood was chosen as the MLE of posterior distribution as shown below:

$\beta = 1.5373$
A = 42.0688
$\gamma_1 = 0.1162$
$\gamma_2 = 0.0059$
$\gamma_3 = 0.0592$

Table 8 summarizes the median of measure of accuracy and correlation of h_{E+Dj} and h_{D39} for the scenarios described above. Also PHM parameters based on statistical data and expert knowledge combined by statistical data for the case having one data history to the case having 39 data histories with their correspondent measure of accuracy are obtained. The detailed results are shown in Appendix B, and are graphically demonstrated in Figures 39 through 47. Figures 39 through 43 compare values of the parameters based on statistical data and with those obtained using both expert knowledge and statistical data. Figures 44 through 47 compare the same cases in terms of values of median for measure of accuracy and correlation between hazards based on the parameters obtained using the two methods.

The results are discussed in section 11.3.6.1

Note: Section 9.1 suggested to us that we use at least 5*100,000 samples when having 100,000 blocks and also that 1 be added to each block. We did not follow this completely because the capability of the current software did not allow us to execute an updating process for so many data. There are recursive procedures that can be used to attain the minimum required samples suggested in section 9.1. However for this case we realized that having a block with 137 sample data removes the need for having a large sample. The reason that in section 9.1 we suggested having at least 5*Q was to make sure that the shape of prior distribution will not change if 1 is added to each block. We suggested adding 1 sample to each block to ensure that no block has zero value. Because, if value of density function for one block is zero in prior distribution it will remain zero during the updating process. Therefore we would miss that block even if it could turn out to be the MLE of the posterior distribution. However as we mentioned before, since there is one block with 137 sample data, the probability of having a MLE candidate of posterior distribution among those blocks with no sample data is virtually zero.

Table 8

PH Parameters Obtained Using Expert Knowledge, Statistical Data, and Both

	E (ME)	E (MLE)	D10 (MLE)	E+D10 (ME)	E+D10 (MLE)	D39 (MLE)	E+D39 (ME)	E+D39 (MLE)
β	1.5593	1.5162	1.2420	1.5562	1.5114	1.8620	1.5778	1.5373
A	41.1004	41.7317	45.8648	41.0780	41.7987	40.4008	42.0972	42.0688
γ_1	0.1214	0.1177	0.1418	0.1217	0.1203	0.0874	0.1190	0.1162
γ_2	0.0070	0.0058	0.0000	0.0069	0.0057	0.0032	0.0070	0.0059
γ_3	0.0540	0.0596	0.0725	0.0539	0.0593	0.0648	0.0568	0.0592
Median of measure of accuracy	1.8054	1.7936	1.9175	1.8573	2.0912	1.0000	1.1591	1.0360
Correlation of h_{E+Dj} and h_{D39}	0.9883	0.9915	0.9854	0.9881	0.9904	0.0000	0.9900	0.9920

Note. E: Expert knowledge
D: Statistical data
D10: First 10 histories of statistical data
ME: Mean estimator (estimates based on mean value of the distribution)
MLE: (maximum likelihood estimator)

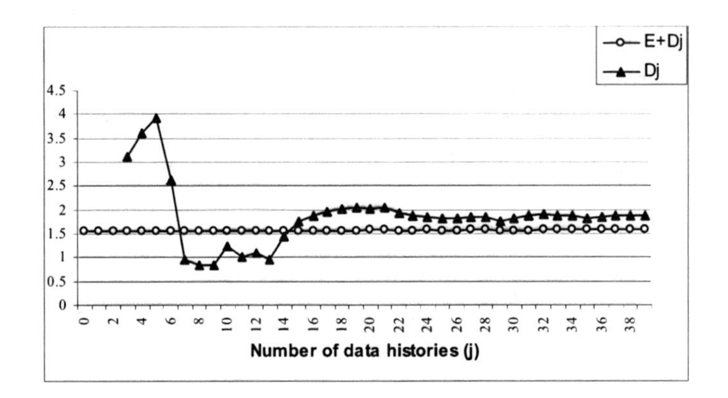

Figure 39: β based on j data histories and β based on knowledge combined with j data histories

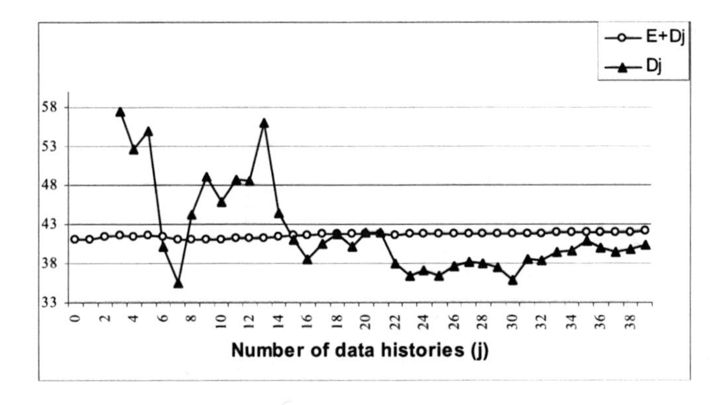

Figure 40: A based on j data histories and A based on knowledge combined with j data histories

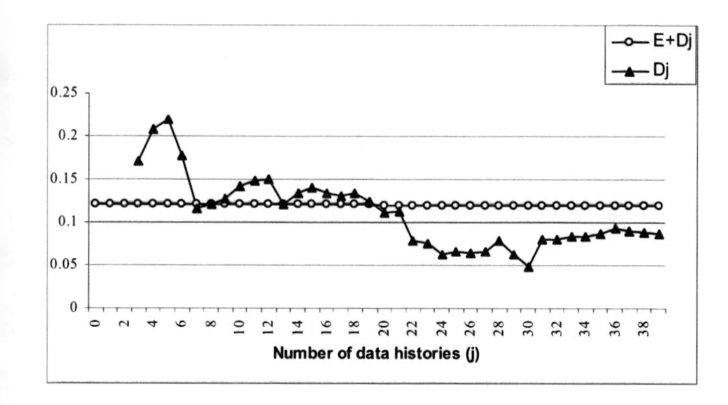

Figure 41: γ_1 **based on j data histories and** γ_1 **based on knowledge combined with j data histories**

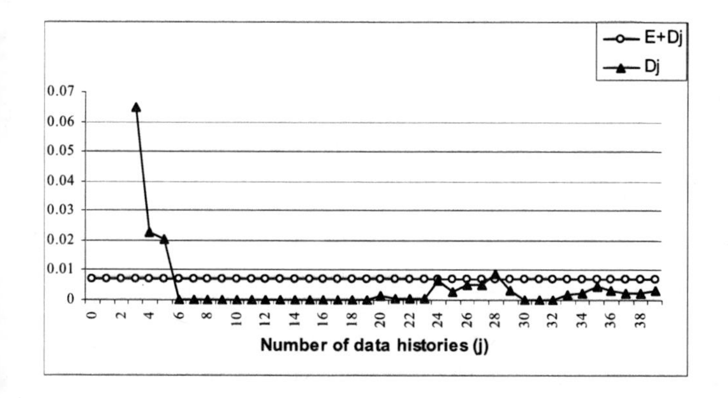

Figure 42: γ_2 **based on j data histories and** γ_2 **based on knowledge combined with j data histories**

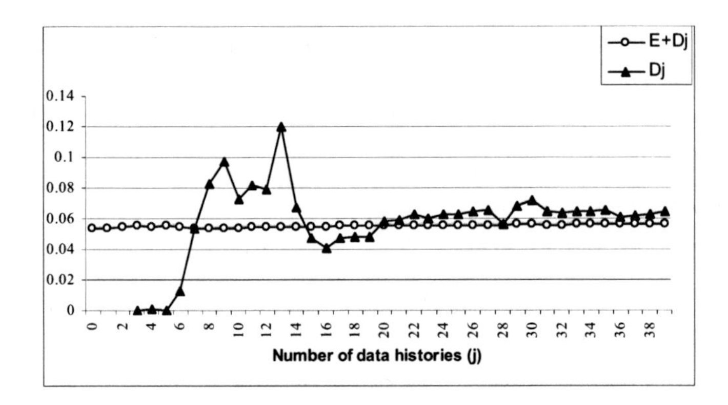

Figure 43: γ_3 based on j data histories and γ_3 based on knowledge combined with j data histories

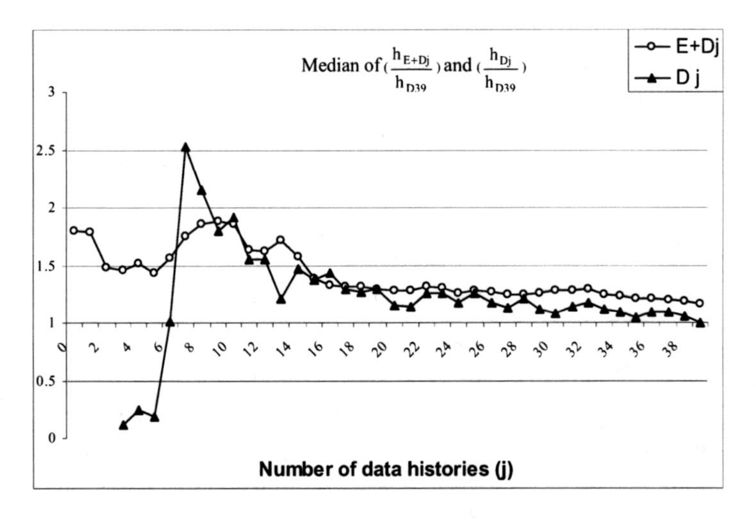

Figure 44: Median of measure of accuracy for PHM based on j data histories and PHM based on knowledge combined with j data histories

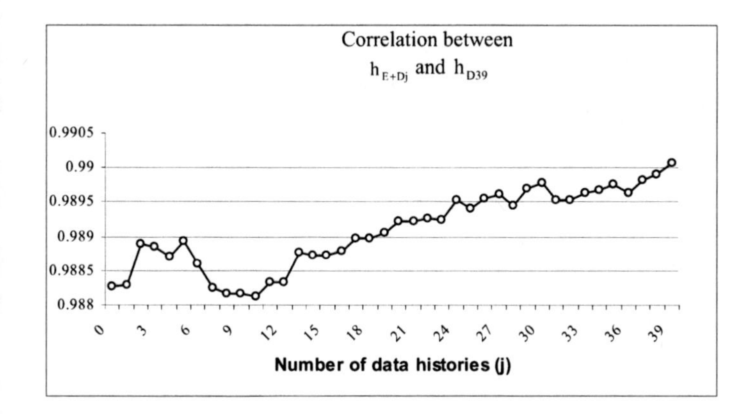

Figure 45: Correlation between hazard based on 39 data histories and the hazard based on knowledge combined with j data histories

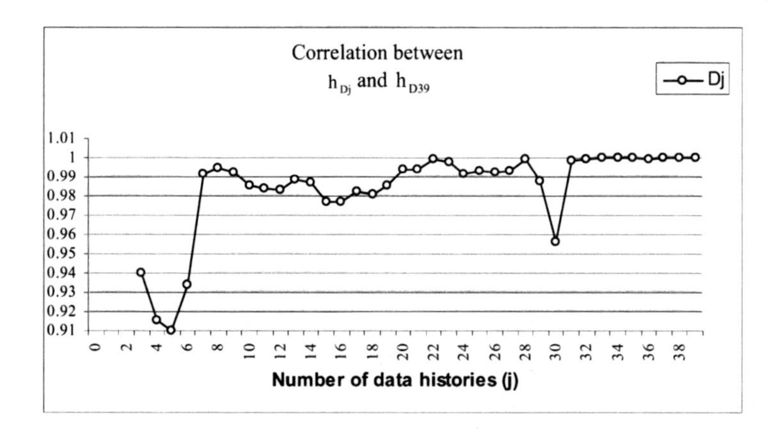

Figure 46: Correlation between hazard based on 39 data histories and the hazard based on j data histories

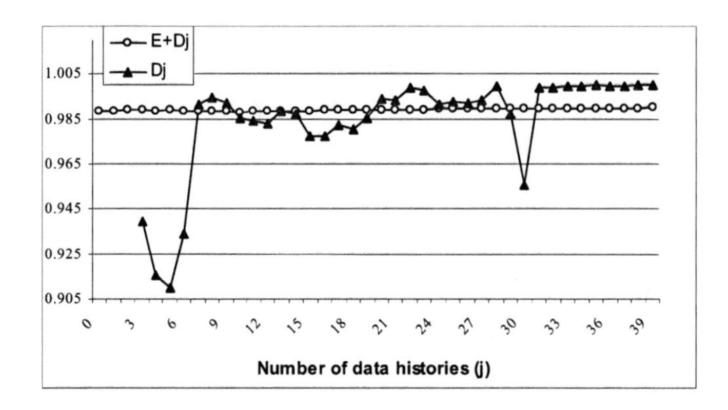

Figure 47: Comparisons between Figures 45 and 46

The updating process was also applied to the results of the experiment after being modified through estimation of the upper and lower bound of the hazard function based on estimation of RUL. The parameters changed very slightly. However more changes can be seen in behavior of measure of accuracy (see Figures 48 though 51).

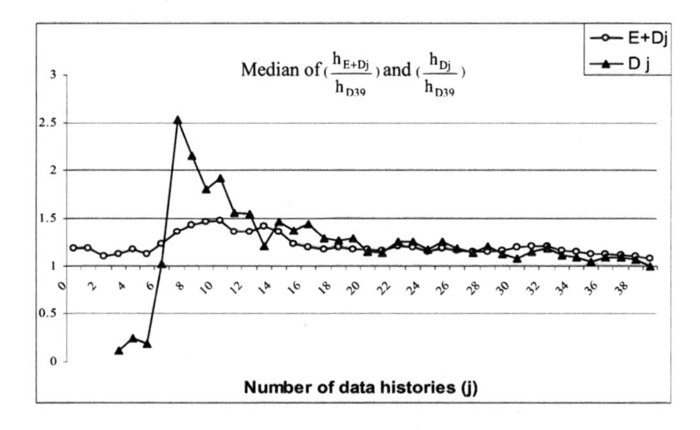

Figure 48: Median of measure of accuracy for PHM based on j data histories and PHM based on knowledge combined with j data histories (2nd try)

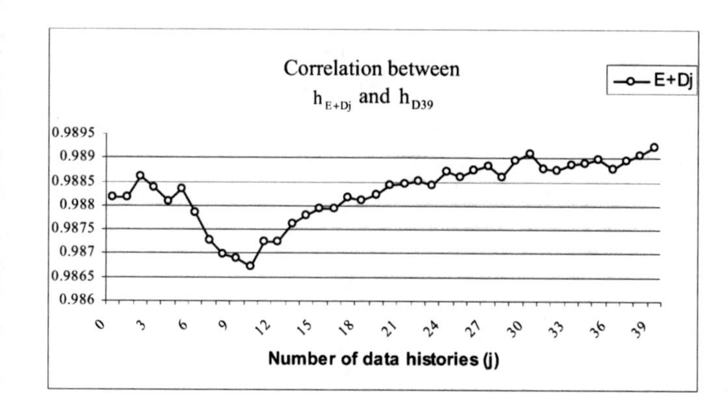

Figure 49: Correlation between hazard based on 39 data histories and the hazard based on knowledge combined with j data histories (2nd try)

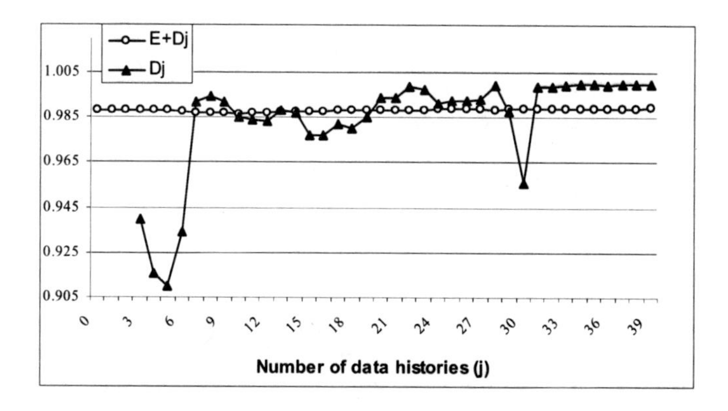

Figure 50: Comparisons between Figures 45 and 49

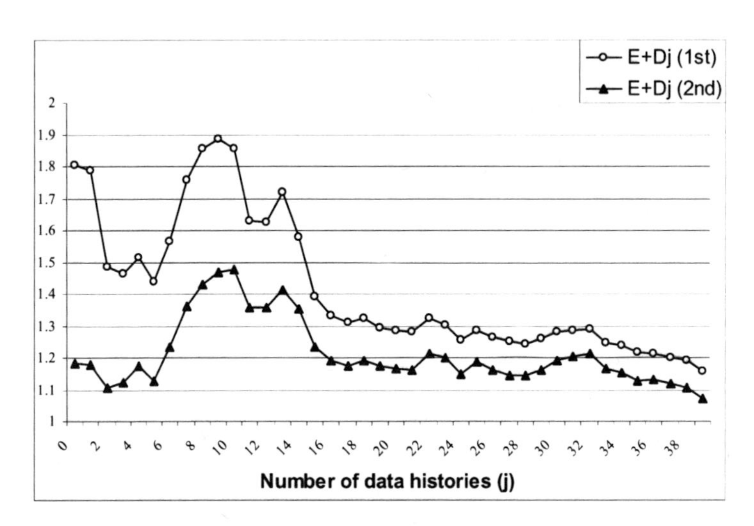

Figure 51: Comparisons between h_{E+Dj} for 1st and 2nd attempts

11.3.6.1. Discussion of the Results of Updating Process

Looking at Figures 39 through 47 and their associated tables in Appendix B, and also Table 8, one can conclude the following (again unless mentioned explicitly, results based on the first attempt included some bias; results based on the second try (attempt)were obtained after removing some of the bias):

1. Maximum likelihood estimators of parameters of PHM are very sensitive to new data when the total number of data histories is small. If we look at the estimated parameters based on maximum likelihood in Figures 39 through 43 we see that they change considerably for the first 30 histories. It is only after 30 histories that the results gradually reach a steady state. For example, MLEs of β starts from 3.1, goes up to 3.9, and then jumps down as low as .83. It has more oscillation with lower amplitude afterwards until it settles after history 25. Parameter A starts at 57, goes down to 35 after few histories, and as for β it oscillates a few more times before it becomes steady. Parameter γ_1 starts at 0.17, increases to 0.219, and its minimum happens at history 30 with a value of 0.048. The magnitude of its maximum value over its minimum value is close to 5 which is very high. Parameter γ_2 begins at 0.065, after two histories it drops to zero, stays zero during about 15 histories, and then fluctuates a few times during the remaining histories. Parameter γ_3 begins at zero, increases to as high as 0.12 during ten histories, and at the end stays around 0.062.
2. Interestingly, when we combine expert knowledge with statistical data we see a very smooth change in values. This may be due to the fact that the prior distribution based on expert knowledge has considerable weight attached to it which prevents estimates of the parameters from changing dramatically.
3. Figure 44 shows that the median of the measure of accuracy follows a pattern similar to those of parameters estimated using maximum likelihood. Median of measure of accuracy starts at 0.11, which implies that MLEs based on the first three histories underestimate the value of the hazard by 9 times.

Following three more histories it suddenly soars to 2.53. After this it gradually decreases to 1 with small fluctuations. Correlation of h_{Dj} and h_{D39} starts at 0.9397 (see Figures 45 through 47), decreases until reaches 0.9100, then gradually improves (getting closer to 1) except at j = 30 where it drops to 0.9559. After this point it approaches 1 smoothly. Median of measure of accuracy and correlation of h_{E+Dj} and h_{D39} for the parameters estimated based on both expert knowledge and statistical data follow the same pattern as those based only on statistical data. However, their fluctuations are much smoother.

4. Based on points 2, 3, and 4 we can say that although values of the parameters estimated based on expert knowledge and statistical data did not change considerably during an updating process, their measure of accuracy showed significant improvement.
5. Figures 44 and 47 also indicate that performance of parameters estimated on pure expert knowledge can be as good as or better than performance of parameters estimated based on 11 data histories. It also shows that combining expert knowledge and statistical data can be beneficial even when we have collected 29 data histories. For Figures 48 through 51 (the results after removing some of the bias) we can say that the accuracy of the hazard based on pure expert knowledge is as good as at least 22 data histories and it is beneficial until the last history of data is combined with it.

12. CONCUSION AND FUTURE RESEARCH

12.1 Conclusion

There has been a continuous effort since the 1970s to combine Bayesian statistics and proportional hazards modeling. Some authors worked with non-parametric PHM while others considered a parametric form of the PHM. Very few authors considered time-varying covariates while most authors assumed that covariates remain constant over data histories. No author has fully addressed parametric PHM with time-dependent covariates. In addition, integration of expert knowledge to estimate parameters of PHM with time-dependent covariates has not been considered in the literature. On the other hand, PHM with time-dependent covariates is being applied in practice in many fields of study especially reliability where most experts do not have deep statistical knowledge. Therefore, a methodology that can extract expert knowledge in a convenient way for both knowledge elicitors and experts is needed.

The knowledge elicitation methodology described in this research does not require special statistical expertise from experts. It is based on case comparisons and case analyses. Therefore, it is easy to understand and experts are comfortable comparing and analyzing cases that describe different machine conditions. For example in the case study carried out in Chaloner et al. (1993), the authors asked the experts to estimate the probability of an event for a period of time under two different treatments (say P_c and P_p). Then they asked the experts to tell them either graphically or parametrically the marginal distribution of the probability of those events (say $f(P_c)$ and $f(P_p)$). Finally, they asked each expert to explain what the joint probability distribution of those estimates would look like (say $f(P_c, P_p)$). In our opinion it is very difficult (or impossible) for an expert to be able to show his/her knowledge regarding the joint probability distribution of two variables, both of which are non-observable quantities. In the Dofasco case study the expert sometimes had problems with simple statistical concepts (simple compared to what authors of Chaloner et al. (1993) asked the experts in their case study) such as confidence intervals. Once when pressed to provide intervals for his estimates based on a given or arbitrary confidence level the Dofasco expert responded:

"... I have conferred with one of our engineers down here as I am still confused about these questions. When you ask about my confidence level, I am basing my answers on our experience, instinct, manpower availability on/off shift, weekends, etc. I don't know that I can always say that I am 90% sure or 70% sure. It is my best guess..."

Based on our experience we believe that although experts in the area of reliability can be a very valuable source of knowledge, they usually have difficulty expressing their knowledge using statistical concepts. The situation may be different in other fields such as medical science. Because of the sensitivities of the knowledge and the results based on that knowledge, experts in the area of medical science may be willing to spend more time learning statistical concepts. However, in the areas of reliability and maintenance, we think it should be the facilitators' responsibility to design questions in such a way that the experts can transfer their knowledge without difficulties.

Any knowledge elicitation process can become very boring and endanger the quality of the extracted knowledge if it lasts for a long time. In our methodology, we expect the knowledge extraction to be performed in five to ten sessions using in total between five to 10 hours of experts' time which seems an affordable period in most cases in today's industrial environment.

The way that the prior distribution is constructed and displayed is effective and efficient in capturing expert knowledge and at the same time is doable using present computational facilities. Usually after 15 to 70 case comparisons, depending on different factors, (also see section 8.3.2) the methodology will be able to build the prior distributions. The total number of case comparisons depends on the number of parameters of a PHM (usually three to six), our expertise in designing cases, and the expert's level of expertise in his/her field.

The performance of the parameters based on pure expert knowledge in the case study addressed in this research was at least as good as the performance of the parameters based on 11 data histories obtained using the method of maximum likelihood. After removing much of the bias in our method the results were improved dramatically which showed that the parameters based on expert knowledge can be as accurate as parameters based on 22 data histories. It was also noted that when statistical data are correlated as is the case in this research, parameters estimated using the method of maximum likelihood and those obtained using either pure expert knowledge or expert knowledge combined with data can be very different and yet can provide very close values for the hazard. Based on this fact we defined an indicator, *measure of accuracy*, which is the ratio of the hazard calculated based on the parameters obtained using a methodology subject to test, to the hazard obtained based on parameters calculated using all statistical data. If the value of this measure of accuracy for a set of statistical data is close to one, we can say that parameters obtained using the methodology subject to test has a similar accuracy to parameters obtained from all statistical data using the method of maximum likelihood. This measure was used throughout this research to evaluate accuracy of parameters estimated using different combinations of expert knowledge and statistical data.

A method for combining knowledge of different experts was suggested. It was mentioned that there are two ways of doing this. Knowledge can be blended during the knowledge elicitation stage using either the Delphi technique or other similar techniques such as behavioral aggregation. Alternatively, we can combine the prior distribution based on knowledge from different experts using the method described in section 9.2.4. However, this technique was not applied in practice because we did not have access to more than one expert.

The updating process is very simple and takes generally between 1 to 10 hours depending on the number of parameters in the model. During the updating process it was revealed that combining expert knowledge with statistical data can be valuable and can increase the accuracy of the model even after collecting 19 data histories in our first attempt which included some bias. After removing the bias, it was shown that combining the expert knowledge and data is of benefit until the last piece of data is collected. It also showed that posterior distribution is resistant to change due to the confidence of the expert when answering the questions. This high confidence has given the prior distribution a significant weight therefore values of the parameters change slowly when statistical data are combined with the expert knowledge using Bayes' rule.

Although these results were obtained in the area of reliability, other fields of study that use PHM such as medical science, finance, and organization demography, can also benefit from the developments accomplished in this research.

12.2 Future Research

The methodology showed promising results; however, it is difficult to make generalization based on one industrial case study. The methodology needs to be tested in more industrial cases and modifications made accordingly. For instance, it can be applied to cases where covariates are different from those in this case study. Here the covariates were pressure and temperature; one could work on cases where covariates included level of contaminants in circulating oil or vibration intensities, to name but two.

Since in this case study we had but one expert, the methodology explained in section 9.2.4 which deals with combining expert knowledge from different sources has not been tested. The case of having a supervisor that can evaluate the expertise of different experts is also a good topic to consider.

The sampling process from a feasible space like the updating process can be time consuming when the total number of parameters exceeds six. When the number of parameters is seven, sampling from a feasible space alone can take four to ten days, as could the updating process. Therefore, developing methodologies that can facilitate both processes will benefit the applicability of this research.

To be able to predict failure of a system or a component, one not only needs accurate estimates for parameters of a PHM but also to know the future behavior of the covariates. Knowing the exact value of the hazard gives a static picture of the system; however, in industry people need to know the future movement of the hazard to be able to take necessary measures in advance. In this research we have not considered the future behavior of the covariates; however it can be considered in future research.

There is also room for those interested in working on how to take into account an expert's uncertainty in the model. As mentioned in section 8.3.3 there are a few ways this can be done, but we have not been able to demonstrate which one works best.

13. REFERENCES

Adler, M., & Ziglio, E. (1996). *Gazing into the oracle: The Delphi method and its application to social policy and public health.* London: Jessica Kingsley.

Anderson, M., Jardine, A. K. S., & Higgins, R. T. (1982). The use of concomitant variables in reliability estimation. *13th Annual Pittsburgh Conference Proceedings* (pp. 73-81). Pittsburgh, PA.

Apostolakis, G. (1978). Probability and risk assessment: The subjectivist viewpoint and some suggestions. *Nuclear Safety, 19*(3), 305-315.

Ascher, H. (1983). Regression analysis of repairable systems reliability. In J. K. Skwirzynski (Ed.), *Electronic systems effectiveness and life cycle costing.* (pp. 15). Berline: Springer.

Autar, R. K. (1996). An automated diagnostic expert system for diesel engines. *Journal of Engineering for Gas Turbines and Power, 118*(3), 673-679.

Badiru, A. B. (1992). *Expert systems applications in engineering and manufacturing.* Upper Saddle River, NJ: Prentice-Hall.

Bainbridge, L. (1999). Verbal reports as evidence of the process operator's knowledge. *International Journal of Human-Computer Studies, 51*(2), 213-238.

Banjevic, D., & Jardine, A. K. S. (2006). Calculation of reliability function and remaining useful life for a Markov failure time process. *IMA Journal of Management Mathematics, 17*(2), 115.

Banjevic, D., Jardine, A. K. S, Makis, V., & Ennis, M. (2001). A control-limit policy and software for condition-based maintenance optimization. *INFOR-Ottawa, 39*(1), 32-50.

Barlow, & Proschan. (1965). *Mathematical theory of reliability.* New York: John Wiley.

Baxter, M. J., Bendell, A., Manning, P. T., & Ryan, S. G. (1988). Proportional hazards modeling of transmission equipment failures. *Reliability Engineering and System Safety, 21*, 129-144.

Beach, L. R., & Swenson, R. G. (1966). Intuitive estimation of means. *Psychonomic Science, 5*, 161–162.

Belkin, N. J., Brooks, H. M., & Daniels, P. J. (1987). Knowledge elicitation using discourse analysis. *International Journal of Man-Machine Studies, 27*(2), 127.

Bell, J., & Hardiman, R. J. (1989). The third role- the naturalistic knowledge engineer. In D. Diaper (Ed.), *Knowledge elicitation: Principles, techniques, and applications.* New York: Halsted Press.

Bendell, A., Walley, M., Wightman, D. W., & Wood, L. M. (1986). Proportional hazards modeling in reliability analysis: An application to brake discs on high speed trains. *Quality and Reliability Engineering International, 2*(1), 45-52.

Ben-Israel, A. (1966). A Newton-Raphson method for the solution of systems of equations. *Journal of Mathematical Analysis and Applications, 15*, 243-252.

Berger, J. O. (1993). *Statistical decision theory and Bayesian analysis.* New York: Springer.

Berger, J. O., & O'Hagan, A. (1988). Ranges of posterior probabilities for unimodal priors with specified quantiles. *Bayesian Statistics, 3*, 45-65.

Berger, P. D., & Maurer, R. E. (2002). *Experimental design with applications in management, engineering and the sciences*. Duxbury/Thomson Learning.

Bernardo, J. M., & Smith, A. F. M. (2001). *Bayesian theory*. Chichester, UK: Wiley.

Besag, J., & Green, P. J. (1993). Spatial statistics and Bayesian computation. *Journal of the Royal Statistical Society, Series B:Methodological, 55*(1), 25-37.

Bloger, F., & Wright, G. (1992). Reliability and validity in expert judgment. In G. Wright, & F. Bolger (Eds.), *Expertise and decision support*. (pp. 30). New York: Plenum.

Bock, R. D., & Jones, L. V. (1968). *The measurement and prediction of judgment and choice*. Holden-Day.

Booker, J., Campbell, K., Goldman, A. G., Johnson, M. E., & Bryson, M. C. (1981). *Application of Cox's proportional hazards model to light water reactor component failure data*. No. LA-8834-SR). Los Alamos, N.M.: Los Alamos Scientific Laboratory.

Bowles, N. (1999). The Delphi technique. *Nursing Standard, 13*(45), 32-36.

Boy, G. (1997). The group elicitation method for participatory design and usability testing. *Interactions, 4*(2), 27.

Bradshaw, J. M., & Boose, J. H. (1990). Decision analysis techniques for knowledge acquisition: Combining information and preferences using Aquinas and Axtol. *International Journal of Man-Machine Studies, 32*(2), 121.

Brenner, L. A., Koehler, D. J., Liberman, V., & Tversky, A. (1996). Overconfidence in probability and frequency judgments: A critical examination. *Organizational Behavior and Human Decision Processes, 65*(3), 212-219.

Breuker, J., & Wielinga, B. (1987). Use of models in the interpretation of verbal data. In A. L. Kidd (Ed.), *Knowledge acquisition for expert systems: A practical handbook*. New York: Plenum Press.

Brooks Stephen P., & Roberts Gareth O. (1998). Convergence assessment techniques for Markov Chain Monte Carlo. *Statistics and Computing, 8*(4), 319-335.

Brown, C. E., & O'Leary, D. E. (1995). Introduction to artificial intelligence and expert systems. *AI/ES Section of the American Accounting Association*, http://www.bus.orst.edu/faculty/brownc/es_tutor/es_tutor.htm#1-AI

Buckley, J. J., & Hayashi, Y. (1993). Hybrid neural nets can be fuzzy controllers and fuzzy expert systems. *Fuzzy Sets and Systems, 60*(2), 135-142.

Bunks, C., & McCarthy, D. (2000). Condition-based maintenance of machines using hidden markov models. *Mechanical Systems and Signal Processing, 14*(4), 597-612.

Campbell, J. D., & Jardine, A. K. S. (Eds.). (2001). *Maintenance excellence: Optimizing equipment life-cycle decisions*. New York: Marcel Dekker.

Campbell, J. D. (1995). *Uptime: Strategies for excellence in maintenance management.* Productivity Press.

Capener, E. J. R., Moyes, A., Burt, G. M., & Mcdonald, J. R. (1995). The application of expert systems to fault diagnosis in alternators. *Nuclear Energy(1978)*, *34*, 53-58.

Carmody, D. P., Kundel, H. L., & Toto, L. C. (1984). Comparison scans while reading chest images: Taught, but not practiced. *Investigative Radiology*, *19*(5), 462.

Carpenter, G. A., Grossberg, S., Markuzon, N., Reynolds, J. H., & Rosen, D. B. (1992). Fuzzy ARTMAP: A neural network architecture for incremental supervised learning of analog multidimensional maps. *IEEE Transactions on Neural Networks*, *3*(5), 698-713.

Chaloner, K. (1996). The elicitation of prior distributions. In D. Berry, & D. Stangl (Eds.), *Bayesian biostatistics*. (pp. 141-156). New York: Marcel Dekker.

Chaloner, K., Church, T., Louis, T. A., & Matts, J. P. (1993). Graphical elicitation of a prior distribution for a clinical trial. *The Statistician*, *42*(4), 341-353.

Chaloner, K., & Rhame, F. S. (2001). Quantifying and documenting prior beliefs in clinical trials. *Statistics in Medicine*, *20*(4), 581-600.

Chee, C., & Power, M. (1990). Expert systems maintainability. *Annual Reliability and Maintainability Symposium Proceedings*, 415-418.

Chen, W., Meher-Homji, C. B., & Mistree, F. (1994). Compromise: An effective approach for condition-based maintenance management of gas turbines. *Engineering Optimization*, *22*, 185-201.

Chinnam, R. B. (2002). On-line reliability estimation for individual components using statistical degradation signal models. *Quality and Reliability Engineering International*, *18*(1), 53-73.

Cleaves, D. A. (1987). Cognitive biases and corrective techniques: Proposals for improving elicitation procedures for knowledge-based systems. *International Journal of Man-Machine Studies*, *27*(2), 155.

Clemen, R. T., Fischer, G. W., & Winkler, R. L. (2000). Assessing dependence: Some experimental results. *Management Science*, *46*(8), 1100-1115.

Coetzee, J. L. (2004). *Maintenance*. Victoria, BC: Trafford Publishing.

Cooke, N. M., & Schvaneveldt, R. W. (1988). Effects of computer programming experience on network representations of abstract programming concept. *International Journal of Man-Machine Studies*, *29*, 407-427.

Cooke, R. M. (1991). *Experts in uncertainty: Opinion and subjective probability in science*. UK: Oxford University Press.

Cooke, R. M., & Goossens, L. H. J. (2000). Procedures guide for structural expert judgment in accident consequence modeling. *Radiation Protection Dosimetry*, *90*, 303-309.

Cooke, N. J. (1994). Varieties of knowledge elicitation techniques. *International Journal of Human-Computer Studies*, *41*(6), 801-849.

Cordingley, E. S. (1989). Knowledge elicitation techniques for knowledge-based systems. In D. Diaper (Ed.), *Knowledge elicitation : Principles, techniques, and applications* (pp. 87). New York: Halsted Press.

Cox, D. R. (1972). Regression models and life tables (with discussion). *Journal of Statistical Society B, 34*, 187-220.

Crawford, J. L., & Weinstock, R. (1990). The NASA trend analysis program. *Proceeding of Annual Reliability Maintenance Systems*, 25-30.

Cullen, J. R., & Bryman, A. (1988). The knowledge acquisition bottleneck: A time for reassessment? *Expert Systems, 5*, 216- 225.

Dawes, R. M. (1979). The robust beauty of improper linear models. *The American Psychologist, 34*, 571.

Diamond, I. D., McDonald, J. W., & Shah, I. H. (1986). Proportional hazards models for current status data: Application to the study of differentials in age at weaning in Pakistan. *Demography, 23*(4), 607-620.

Diederich, J., Ruhmann, I., & May, M. (1987). KRITON: A knowledge-acquisition tool for expert systems. *International Journal of Man-Machine Studies, 26*(1), 29.

Drake, P. R., Jennings, A. D., Grosvendor, R. I., & Whittleton, D. (1995). A data acquisition system for machine tool condition monitoring. *Quality and Reliability Engineering International, 11*, 15-26.

Drury, C. G. (1990). Methods for direct observation of performance. In J. R. Wilson, & E. N. Corlett (Eds.), *Evaluation of human work : A practical ergonomics methodology*. New York: Taylor & Francis.

Ekman, P., & Scherer, K. R. (1982). *Handbook of methods in nonverbal behavior research*. Paris: Editions de la Maison des Sciences de l'Homme.

Elstein, A. S., Shulman, L. S., & Sprafka, S. A. (1978). Medical problem solving: An analysis of clinical reasoning. Cambridge, MA: Harvard University Press.

Ericsson, K. A. (1984). *Protocol analysis: Verbal reports as data*. Cambridge, MA: MIT Press.

Erlick, D. E. (1964). Absolute judgments of discrete quantities randomly distributed over time. *Journal of Experimental Psychology, 67*, 475-482.

Faraggi, D., & Simon, R. (1997). Large sample Bayesian inference on the parameters of the proportional hazard models. *Statistics in Medicine, 16*(22), 2573-2585.

Farewell, V. T. (1979). An application of Cox's proportional hazard model to multiple infection data. *Applied Statistics, 28*(2), 136-143.

Ferguson, T. S. (1973). A Bayesian analysis of some non-parametric problems. *The Annals of Statistics, 1*(2), 209-230.

Fischhoff, B., & Beyth, R. (1975). I knew it would happen: Remembered probabilities of once-future things. *Organizational Behavior and Human Performance, 13*(1), 1–16.

Fischhoff, B., Slovic, P., & Lichtenstein, S. (1978). Fault trees: Sensitivity of assessed failure probabilities to problem representation. *Journal of Experimental Psychology: Human Perception and Performance, 4*, 330-344.

French, S. (1985). Group consensus probability distributions: A critical survey. *Bayesian Statistics, 2*, 183-202.

Gale, K. W., & Watton, J. (1999). A real-time expert system for the condition monitoring of hydraulic control systems in a hot steel strip finishing mill. *Proceedings of the Institution of Mechanical Engineers, Part I: Journal of Systems and Control Engineering, 213*(5), 359-374.

Gammack, J. G. (1987). Different techniques and different aspects on declarative knowledge. In A. L. Kidd (Ed.), *Knowledge acquisition for expert systems: A practical handbook*. New York: Plenum Press.

Garthwaite, P. H. (2005). Statistical methods for eliciting probability distributions. *Journal of the American Statistical Association, 100*(470), 680.

Garthwaite, P. H., & Dickey, J. M. (1988). Quantifying expert opinion in linear regression problems. *Journal of the Royal Statistical Society, Series B: Methodological, 50*(3), 462-474.

Garthwaite, P. H., Kadane, J. B., & O'Hagan, A. (2005). Statistical methods for eliciting probability distributions. *Journal of the American Statistical Association, 100*(470), 680-701.

Garthwaite, P. H., & O'Hagan, A. (2000). Quantifying expert opinion in the UK water industry: An experimental study. *The Statistician, 49*(4), 455-477.

Gasmi, S., Love, C. E., & Kahle, W. (2003). A general repair, proportional-hazards, framework to model complex repairable systems. *IEEE Transactions on Reliability, 52*(1), 26-32.

Gelfand, A. E., & Smith, A. F. M. (1990). Sampling-based approaches to calculating marginal densities. *Journal of the American Statistical Association, 85*(410), 398-409.

Gelfand, A. E., & Mallick, B. K. (1995). Bayesian analysis of proportional hazards models built from monotone functions. *Biometrics, 51*(3), 843-852.

Gelgele, H. L., & Wang, K. (1998). An expert system for engine fault diagnosis: Development and application. *Journal of Intelligent Manufacturing, 9*(6), 539-545.

Gelman, A., Carlin, J. B., Stern, H. S., & Rubin, D. B. (2004). *Bayesian data analysis*. Florida: Chapman & Hall/CRC.

Geman, S., & Geman, D. (1984). Stochastic relaxation, Gibbs distributions, and the Bayesian restoration of images. *IEEE Transactions on Pattern Analysis and Machine Intelligence, 6*(6), 721–741.

Genest, C., & Zidek, J. V. (1986). Combining probability distributions: A critique and an annotated bibliography. *Statistical Science, 1*(1), 114-135.

Gescheider, G. A. (1997). *Psychophysics: The fundamentals*. Lawrence Erlbaum.

Ghorai, J. K. (1989). Non-parametric Bayesian estimation of a survival function under the proportional hazard model. *Communication in Statistics, 18*, 1831—1842.

Gill, J., & Walker, L. D. (2005). Elicited priors for Bayesian model specifications in political science research. *Journal of Politics*, *67*(3), 841.

Gokhale, D. V., & Press, S. J. (1982). Assessment of a prior distribution for the correlation coefficient in a bivariate normal distribution. *Journal of the Royal Statistical Society, Series A :General*, *145*(2), 237-249.

Goldstein, M. (1999). Bayes' linear analysis. *Encyclopaedia of Statistical Sciences, Update*, *3*, 29–34.

Gore, S. M., Pocock, S. J., & Kerr, G. R. (1984). Regression models and non-proportional hazards in the analysis of breast cancer survival. *Applied Statistics*, *33*(2), 176-195.

Graesser, A. C., & Murray, K. (1990). A question-answer methodology for exploring a user's acquisition and knowledge of a computer environment. In J. B. Black, S. P. Robertson & W. Zachary (Eds.), (Cognition, computing, and cooperation. ed.) (pp. 30). Norwood, NJ: Ablex.

Graesser, A. C., & Clark, L. F. (1985). *Structures and procedures of implicit knowledge*. Norwood, NJ: Ablex.

Gray, C., Harris, N., Bendell, A., & Walker, E. (1988). The reliability analysis of weapons systems. *Reliability Engineering and System Safety*, *21*(4), 245-70.

Gray, R. J. (1994). A Bayesian analysis of institutional effects in a multicenter cancer clinical trial. *Biometrics*, *50*(1), 244-253.

Grenander, U. (1983). *Tutorial in pattern theory*. No. Lecture Notes). Brown University: Division of Applied Mathematics.

Group Paper. (1998). Discussion on papers on 'Elicitation'. *The Statistician*, *47*(1), 55-68.

Hammersley, J. M., & Handscomb, D. C. (1979). *Monte Carlo methods*. London: Chapman and Hall.

Hampton, J. M., Moore, P. G., & Thomas, H. (1973). Subjective probability and its measurement. *Journal of the Royal Statistical Society, Series A :General*, *136*(1), 21-42.

Harrison, N. (1995). Oil condition monitoring for the railway business. *Insight*, *37*, 278-283.

Hastings, W. K. (1970). Monte Carlo sampling methods using Markov chains and their applications. *Biometrika*, *57*(1), 97.

Haykin, S. (1994). *Neural networks, A comprehensive foundations*. New York: Maxwell Macmillan.

He, Z., Wu, M., & Gong, B. (1992). Neural network and its application on machinery fault diagnosis. *IEEE International Conference on Systems Engineering*, 576-579.

Henebry, K. L. (1996). Do cash flow variables improve the predictive accuracy of a Cox proportional hazards model for bank failure? *The Quarterly Review of Economics and Finance*, *36*(3), 395-409.

Hink, R. F., & Woods, D. L. (1987). How humans process uncertain knowledge: An introduction for knowledge engineers. *AI Magazine*, *8*(3), 41.

Hinton, G. E. (1992). How neural networks learn from experience. *Scientific American*, *267*(3), 144-151.

Hjort, N. L. (1990). Non-parametric Bayes' estimators based on beta processes in models for life history data. *The Annals of Statistics, 18*(3), 1259-1294.

Hoffman, R. R. (1987). The problem of extracting the knowledge of experts. *AI Magazine, 8*, 53-67.

Hogarth, R. M. (1987). *Judgment and choice: The psychology of decision*. Chichester, UK: Wiley.

Hogarth, R. M. (1975). Cognitive processes and the assessment of subjective probability distributions. *Journal of the American Statistical Association, 70*(350), 271-289.

Holtzman, S. (1988). *Intelligent decision systems*. Boston, MA: Addison-Wesley Longman.

Hopfield, J. J. (1982). Neural networks and physical systems with emergent collective computational abilities. *Proceedings of the National Academy of Sciences, 79*(8), 2554-2558.

Horstkotte, E. (2000). *Fuzzy expert systems*.http://www.austinlinks.com/Fuzzy/expert-systems.html

Huber, G. P. (1974). Methods for quantifying subjective probabilities and multi-attribute utilities. *Decision Sciences, 5*(3), 430.

Ibrahim, J. G., Chen, M. H., Chen, M. H., & Sinha, D. (2001). *Bayesian survival analysis*. New York: Springer.

Ibrahim, J. G., & Laud, P. W. (1994). A predictive approach to the analysis of designed experiments. *Journal of the American Statistical Association, 89*(425), 309-319.

Irony, T. Z., & Singpurwalla, N. D. (1997). Noninformative priors do not exist: A discussion with Jose M. Bernardo. *Journal of Statistical Interference and Planning, 65*(1), 159-189.

Jardine, A. K. S. (2002). Optimizing condition-based maintenance decisions. *Annual Reliability and Maintainability Symposium Proceedings*, Seattle, WA.

Jardine, A. K. S., & Anderson, M. (1985). Use of concomitant variables for reliability estimation. *Maintenance Management International, 5*(2), 135-140.

Jardine, A. K. S., Anderson, P. M., & Mann, D. S. (1987). Application of the Weibull proportional hazards model to aircraft and marine engine failure data. *Quality and Reliability Engineering International, 3*(2), 77-82.

Jardine, A. K. S., Banjevic, D., & Makis, V. (1997). Optimal replacement policy and the structure of software for condition-based maintenance. *Journal of Quality in Maintenance Engineering, 3*(2), 109-119.

Jardine, A. K. S., Ralston, P., Reid, N., & Stafford, J. (1989). Proportional hazards analysis of diesel engine failure data. *Quality and Reliability Engineering International, 5*(3), 207-216.

Jardine, A. K. S., & Tsang, A. H. C. (2006). *Maintenance, replacement, and reliability : Theory and applications*. Boca Raton, FL: CRC/Taylor & Francis.

Javadpour, R., & Knapp, G. M. (2003). A fuzzy neural network approach to machine condition monitoring. *Computer Science and Industrial Engineering, 45*, 323.

Jeffreys, H. (1961). *Theory of probability*. (3rd ed.). Oxford: Clarendon Press.

Johnson, E. J. J., Hershey, J. J., Meszaros, J. J., & Kunreuther, H. J. (1993). Framing, probability distortions, and insurance decisions. *Journal of Risk and Uncertainty*, *7*(1), 35-51.

Johnson, L., & Johnson, N. (1987). Knowledge elicitation involving teachback interviewing. In A. L. Kidd (Ed.), *Knowledge acquisition for expert systems : A practical handbook*. New York: Plenum Press.

Johnson, N. L., & Kotz, S. (1972). *Distributions in statistics: Continuous multivariate distributions*. New York: Wiley.

Jolliffe, I. T. (1986). *Principal component analysis*. New York: Springer.

Kabus, I. (1976). You can bank on uncertainty. *Harvard Business Review*, *56*, 95-105.

Kadane, J. B. (1980). Predictive and structural methods for eliciting prior distributions. In A. Zellner (Ed.), *Bayesian analysis in econometrics and statistics* (pp. 89–93). Amsterdam: North-Holland.

Kadane, J. B., Dickey, J. M., Winkler, R. L., Smith, W. S., & Peters, S. C. (1980). Interactive elicitation of opinion for a normal linear model. *Journal of the American Statistical Association*, *75*(372), 845-854.

Kadane, J. B., & Schum, D. A. (1996). *A probabilistic analysis of the sacco and vanzetti evidence*. New York: Wiley.

Kadane, J. B., & Winkler, R. L. (1988). Separating probability elicitation from utilities. *Journal of the American Statistical Association*, *83*(402), 357-363.

Kadane, J. B., & Wolfson, L. J. (1998). Experiences in elicitation. *The Statistician*, *47*(1), 3-19.

Kahneman, D., & Tversky, A. (1973). On the psychology of prediction. *Psychological Review*, *80*(4), 237-251.

Kalbfleisch, J. D. (1978). Non-parametric Bayesian analysis of survival time data. *Journal of the Royal Statistical Society, Series B :Methodological*, *40*(2), 214-221.

Kawahara, K., Sasaki, H., Kubokawa, J., Asahara, H., & Sugiyama, K. (1998a). A proposal of a supporting expert system for outage planning of electric power facilities retaining high power supply reliability. II. knowledge processing and simulation results – part I – outline of a supporting system and outage work allocation based on indices. *IEEE Transactions on Power Systems*, *13*(4), 1453-1458.

Kawahara, K., Sasaki, H., Kubokawa, J., Asahara, H., & Sugiyama, K. (1998b). A proposal of a supporting expert system for outage planning of electric power facilities retaining high power supply reliability. II. knowledge processing and simulation results – part II - knowledge processing and simulation results. *IEEE Transactions on Power Systems*, *13*(4), 1459-1465.

Keeney, R. L., & Raiffa, H. (1993). *Decisions with multiple objectives: Preferences and value trade-offs*. UK: Cambridge University Press.

Kelly, A. (2003). *Maintenance organization and systems*. Oxford, UK: Elsevier Science.

Keren, G. (1992). Improving decisions and judgments: The desirable versus the feasible. In G. Wright & F. Bolger (Eds.), *Expertise and decision support* (pp. 22). New York: Plenum.

Kim, Y., & Lee, J. (2003). Bayesian analysis of proportional hazard models. *Annals of Statistics, 31*(2), 493-511.

Kim, S. W., & Ibrahim, J. G. (2000). On Bayesian inference for proportional hazards models using noninformative priors. *Lifetime Data Analysis, 6*(4), 331-341.

Klein, M., & Methlie, L. B. (1990). *Expert systems: A decision support approach.* Workingham, UK: Addison-Wesley.

Knapp, G. M., & Wang, B. H. S. (1992). Machine fault classification: A neural network approach. *International Journal of Production Research, 30*(4), 811-823.

Knapp, G. M., Javadpour, R., & Wang, H. P. (2000). An ARTMAP neural network-based machine condition monitoring system. *Journal of Quality in Maintenance Engineering, 6*(2), 86-105.

Kobbacy, K. A. H., Fawzi, B. B., & Percy, D. F. (1997). A full history proportional hazards model for preventive maintenance scheduling. *Quality and Reliability Engineering International, 13*(4), 187-198.

Krivtsov, V. V., Tananko, D., & Davis, T. (2002). Regression approach to tire reliability analysis. *Reliability Engineering and System Safety, 78*(3), 267-273.

Kuehl, R. O. (2000). *Design of experiments: Statistical principles of research design and analysis.* Duxbury/Thomson Learning.

Kumar, D., & Klefsjo, B. (1994). Proportional hazards model: A review. *Reliability Engineering and System Safety, 44*(2), 177.

Kumar, D., Klefsjoe, B., & Kumar, U. (1992). Reliability analysis of power transmission cables of electric mine loaders using the proportional hazards model. *Reliability Engineering & System Safety, 37*(3), 217-222.

Kumar, D., & Westberg, U. (1997). Maintenance scheduling under age replacement policy using proportional hazards model and TTT-plotting. *European Journal of Operational Research, 99*(3), 507-515.

Kumar, D., & Westberg, U. (1996). Proportional hazards modeling of time-dependent covariates using linear regression: A case study [mine power cable reliability]. *IEEE Transactions on Reliability, 45*(3), 386-392.

Kumara, S. R., Kashyap, R. L., & Soyster, A. L. (1989). Artificial intelligence and manufacturing: An introduction. *Artificial Intelligence: Manufacturing Theory and Practice.*,

Lane, W. R., Looney, S. W., & Wansley, J. W. (1986). An application of the Cox proportional hazards model to bank failure. *Journal of Banking and Finance, 10*(4), 511–531.

Lathrop, R. G. (1967). Perceived variability. *Journal of Experimental Psychology, 73*(4), 498-502.

Laud, P. W., Damien, P., & Smith, A. F. M. (1998). Bayesian non-parametric and covariate analysis of failure time data. *Practical Non-Parametric and Semi Parametric Bayesian Statistics, Lecture Notes in Statistics, 133*, 213-225.

Lee, J. H., & Yang, S. H. (2001). Fault diagnosis and recovery for a CNC machine tool thermal error compensation system. *Journal of Manufacturing Systems, 19*(6), 428-434.

Lewis, S. (1991). Cluster analysis as a technique to guide interface design. *International Journal of Man-Machine Studies, 35*(2), 251.

Li, C. J., & Li, S. Y. (1995). Acoustic emission analysis for bearing condition monitoring. *Wear, 185*, 67-74.

Lichtenstein, S., & Fischhoff, B. (1980). Training for calibration. *Organizational Behavior and Human Performance, 26*(2), 149-171.

Lichtenstein, S., Fischhoff, B., Phillips, L. D., Kahneman, D., Slovic, P., & Tversky, A. (1982). Calibration of probabilities: The state of the art to 1980. In D. Kahneman, P. Slovic & A. Tversky (Eds.), *Judgment under uncertainty: Heuristics and biases* (pp. 306-334). Cambridge: Cambridge University Press.

Likert, R. A. (1932). A technique for measurement of attitudes. *Archives of Psychology, 140*, 5.

Lin, H., Yih, Y., & Salvendy, G. (1995). Neural-network based fault diagnosis of hydraulic forging presses in china. *International Journal of Production Research, 33*(7), 1939-1951.

Lindley, D. V., & Smith, A. F. M. (1972). Bayes' estimates for the linear model. *Journal of the Royal Statistical Society, Series B: Methodological, 34*(1), 1-41.

Lindqvist, B., Molnes, E., & Rausand, M. (1988). Analysis of SCSSV performance data. *Reliability Engineering and System Safety, 20*(1), 3-17.

Liptrot, D., & Palarchio, G. (2000). Utilizing advanced maintenance practices and information technology to achieve maximum equipment reliability. *International Journal of Quality & Reliability Management, 17*(8), 919-928.

Love, C. E., & Guo, R. (1991). Application of Weibull proportional hazards modeling to bad-as-old failure data. *Quality and Reliability Engineering International, 7*(3), 149-157.

Luxhoslashj, J., & Williams, T. (1996). Integrated decision support for aviation safety inspectors. *Finite Elements in Analysis and Design, 23*(2), 381-403.

Makis, V., & Jardine, A. K. S. (1992). Optimal replacement in the proportional hazards model. *INFOR: Information Systems & Operational Research, 30*(2).

Makis, V., & Jardine, A. K. S. (1991a). Computation of optimal policies in replacement models. *IMA Journal of Management Mathematics, 3*(3), 169.

Makis, V., & Jardine, A. K. S. (1991b). Optimal replacement of a production system-a proportional hazards model. *Proceedings of the XIth International Conference on Production Research*, Hefei, China. 718-722.

Mandi, A. M. (1988). The role of human factors in expert systems design and acceptance. *Human Factors, 30*(4), 395.

McCall, J. J. (1965). Maintenance policies for stochastically failing equipment: A survey. *Management Science, 11*, 493-524.

McConway, K. J. (1981). Marginalization and linear opinion pools. *Journal of the American Statistical Association, 76*(374), 410-414.

McCrea, T. (1992). The shuttle processing contractors (SPC) reliability program at the Kennedy space center – the real world. *Proceedings Annual Reliability and Maintainability Symposium*, 376-378.

Meeker, W. Q., & Escobar, L. A. (1998). *Statistical methods for reliability data.* Wiley.

Meher-Homji, C. B., Mistree, F., & Karadikkar, S. (1994). An approach for the integration of condition monitoring and multi-objective optimization for gas turbine maintenance management. *International Journal of Turbo and Jet Engines, 11*, 43-51.

Meister, D. (1989). *Conceptual aspects of human factors.* Baltimore, MD: Johns Hopkins University Press.

Merrick, J. R. W., Soyer, R., & Mazzuchi, T. A. (2003). A Bayesian semiparametric analysis of the reliability and maintenance of machine tools. *Technometrics, 45*(1), 58-69.

Meyer, M. A., & Booker, J. M. (2001). *Eliciting and analyzing expert judgment.* Philadelphia, PA: Society for Industrial and Applied Mathematics and American Statistical Association.

Milgram, P. (2003). *Introduction to questionnaire design (lecture notes for course MIE 1403F, methods in human factors research*). Unpublished manuscript.

Miller, G. A. (1969). A psychological method to investigate verbal concepts. *Journal of Mathematical Psychology, 6*, 169.

Milne, R. (1996). Continuous expert diagnosis: Is the future so far away? *Fuel and Energy Abstracts, 37*(1), 64-64.

Mitra, S., & Pal, S. K. (1995). Fuzzy multi-layer perception, inferencing and rule generation. *IEEE Transactions on Neural Networks, 6*(1), 51-63.

Miyamoto, S., Oi, K., Abe, O., Katsuya, A., & Nakayama, K. (1986). Directed graph representations of association structures: A systematic approach. *IEEE Transactions on Systems, Man, and Cybernetics, 16*(1), 53.

Montgomery, D. C. (2005). *Introduction to statistical quality control.* New York: Wiley.

Morgan, M. G., & Henrion, M. (1990). *Uncertainty: A guide to dealing with uncertainty in quantitative risk and policy analysis.* Cambridge, UK: Cambridge University Press.

Moubray, J. (1997). *Reliability-centered maintenance* (2nd ed.). New York: Industrial Press.

Murphy, A. H., & Winkler, R. L. (1974). Credible interval temperature forecasting- some experimental results. *Monthly Weather Review, 102*(11), 784-794.

Murphy, A. H., & Winkler, R. L. (1977). Can weather forecasters formulate reliable forecasts of precipitation and temperature. *National Weather Digest, 2*, 2-9.

Nash, H. (1964). The judgment of linear proportions. *The American Journal of Psychology, 77*, 480-484.

Nisbett, R. E., Borgida, E., Crandall, R., & Reed, H. (1976). Popular induction: Information is not necessarily informative. In L. Erlbaum Associates (Ed.), *Cognition and social behavior* (pp. 113–133). New York: Hillsdale.

Nyman, D., & Levitt, J. (2001). *Maintenance planning, scheduling, and coordination.* New York: Industrial Press Inc.

O'Hagan, A., & Oakley, J. E. (2004). Probability is perfect, but we can't elicit it perfectly. *Reliability Engineering and System Safety, 85*, 239–248.

Oakley, J., & O'Hagan, A. (2002). Uncertainty in prior elicitations: A non-parametric approach. *Department of Probability and Statistics, Research Report, 521*(02)

Oberkampf, W. L., Helton, J. C., Joslyn, C. A., Wojtkiewicz, S. F., & Ferson, S. (2004). Challenge problems: Uncertainty in system response given uncertain parameters. *Reliability Engineering and System Safety, 85*(1), 11.

Ogaji, S. O. T., & Singh, R. (2002). Advanced engine diagnostics using artificial neural networks. *IEEE International Conference on Artificial Intelligence Systems*, 236-241.

O'Hagan, A. (1998). Eliciting expert beliefs in substantial practical applications. *The Statistician, 47*(1), 21-35.

Oien, K. (1998). Improved quality of input data for maintenance optimization using expert judgment. *Reliability Engineering and System Safety, 60*(2), 93-101.

Olson, J. R., & Rueter, H. H. (1987). *Extracting expertise from experts: Methods for knowledge acquisition.* Ann Arbor, MI: University of Michigan.

Oman, S. D. (1985). Specifying a prior distribution in structured regression problems. *Journal of the American Statistical Association, 80*(389), 190-195.

Oslon, J. R., & Bioslsi, K. J. (1991). Techniques for representing expert knowledge. In K. A. Ericsson, & J. Smith (Eds.), *Toward a general theory of expertise : Prospects and limits* (pp. 46). New York: Cambridge University Press.

Parry, G., & Winter, P. (1981). Characterization and evaluation of uncertainty in probabilistic risk analysis. *Nuclear Safety, 22*(1), 28-42.

Patro, S., & Kolarik, W. J. (1997). Neural networks and evolutionary computation for real-time quality control of complex processes. *Proceedings of the Annual Reliability and Maintainability Symposium*, 327-332.

Percy, D. F. (2004). Subjective priors for maintenance models. *Journal of Quality in Maintenance Engineering, 10*(3), 221-227.

Percy, D. F. (2002). Bayesian enhanced strategic decision-making for reliability. *European Journal of Operational Research, 139*(1), 133-145.

Peterson, C. R., Snapper, K. J., & Murphy, A. H. (1972). Credible interval temperature forecasts. *Bulletin of the American Meteorological Society, 53*(10), 966-1073.

Peterson, C., & Miller , A. (1964). Mode, median, and mean as optimal strategies. *Journal of Experimental Psychology, 68*, 363-367.

Peterson, C. R., & Beach, L. R. (1967). Man as an intuitive statistician. *Psychological Bulletin, 68*(1), 29-46.

Philips, L. D. (1999). *Group elicitation of probability distributions: Are many heads better than one?*. Boston: Klumer Academic.

Pill, J. (1970). *The Delphi method: Substance, context, a critique, and an annotated bibliography.* Case Western Reserve University.

Pitz, G. F. (1965). Response variables in the estimation of relative frequency. *Perceptual and Motor Skills, 21*(3), 867-873.

Rao, B. K. N. (1996). The need for condition monitoring and maintenance management in industries. In Rao, B. K. N. (Ed.), *Handbook of condition monitoring* (pp. 1-37). Oxford, UK: Elsevier Science Ltd.

Riley, G. (2004). *What are expert systems?*http://www.ghg.net/clips/ExpertSystems.html

Sarle, W. S. (2002). *Neural networks FAQ: Introduction.* ftp://ftp.sas.com/pub/neural/FAQ.html

Sasranga, H., & Knezevic, J. (2001). Reliability prediction for condition based maintained systems. *Reliability Engineering and System Safety, 71*, 219-224.

Scarf, P. A. (1997). On the application of mathematical models in maintenance. *European Journal of Operation Research, 99*, 493-506.

Schaefer, R. E., & Borcherding, K. (1973). The assessment of subjective probability distributions: A training experiment. *Acta Psychologica, 37*(2), 117-129.

Schmuller, J. (1992). Expert systems: A quick tutorial. *Journal of Information Systems Education, 9*(4), 92.

Schweickert, R., Burton, A. M., Taylor, N. K., Corlett, E. N., Shadbolt, N. R., & Hedgecock, A. P. (1987). Comparing knowledge elicitation techniques: A case study. *Artificial Intelligence Review, 1*(4), 245 - 253.

Seaver, D. A., & Stillwell, W. G. (1983). *Procedures for using expert judgment to estimate human error probabilities in nuclear power plant operations.* No. NUREG/CR-2743, SAND82-7054). Albuquerque, NM: Sandia National Laboratories.

Shadolt, N., & Burton, M. (1990). Knowledge elicitation: A systematic approach. In J. R. Wilson, & E. N. Corlett (Eds.), *Evaluation of human work : A practical ergonomics methodology* (pp. 406-441). New York: Taylor & Francis.

Shanteau, J. (1992). The psychology of experts: An alternative view. In G. Wright, & F. Bolger (Eds.), *Expertise and decision support*. New York: Plenum.

Shaw, M. L. G., & Gianes, B. R. (1987). An interactive knowledge elicitation technique using personal construct technology. In A. L. Kidd (Ed.), *Knowledge acquisition for expert systems: A practical handbook*. New York: Plenum.

Shepard, R. N. (1962a). The analysis of proximities: Multidimensional scaling with an unknown distance function I. *Psychometrika, 27*(2), 125.

Shepard, R. N. (1962b). The analysis of proximities: Multidimensional scaling with an unknown distance function II. *Psychometrika, 27*(2), 219.

Shuford, E. H. (1961). Percentage estimation of proportion as a function of element type, exposure type, and task. *Journal of Experimental Psychology, 61*, 430–436.

Shyur, H. J., Elsayed, E., & Luxhoj, J. (1999). A general hazard regression model for accelerated life testing. *Annals of Operations Research, 91*, 263-280.

Simpson, W., & Voss, J. F. (1961). Psychophysical judgments of probabilistic stimulus sequences. *Journal of Experimental Psychology, 62*, 416-422.

Sinclair, M. A. (1995). Subjective assessment. In J. R. Wilson, & E. N. Corlett (Eds.), *Evaluation of human work: A practical ergonomics methodology*. (pp. 69-101). London: Taylor & Francis.

Singpurwalla, N. D., & Song, M. S. (1988). Reliability analysis using Weibull lifetime data and expert opinion. *IEEE Transactions on Reliability, 37*(3), 340-347.

Singpurwalla, N., & Song, M. (1987). The analysis of Weibull lifetime data incorporating expert opinion. In R. Viertl (Ed.), *Probability and Bayesian statistics*. New York: Plenum Publishing Corporation.

Siu, N. O., & Kelly, D. L. (1998). Bayesian parameter estimation in probabilistic risk assessment. *Reliability Engineering and System Safety, 62*(1), 89-116.

Slovic, P., & Lichtenstein, S. (1971). Comparison of Bayesian and regression approaches to the study of information processing in judgment. *Organizational Behavior and Human Performance, 6*(6), 649-744.

Spencer, J. (1963). A further study of estimating averages. *Ergonomics, 6*, 255–265.

Spencer, J. (1961). Estimating averages. *Ergonomics, 4*, 317-328.

Stevens, S. S., & Galanter, E. H. (1957). Ratio scales and category scales for a dozen perceptual continua. *Journal of Experimental Psychology, 54*(6), 377-411.

Stillwell, W. G., Seaver, D. A., & Schwartz, J. P. (1982). *Expert estimation of human error probabilities in nuclear plant operations: A review of probability assessment and scaling*. No. NUREG/CR-2255, SAND81-7140). Albuquerque, NM: Sandia National Laboratories.

Tiao, G. C., & Zellner, A. (1964). Bayes' theorem and the use of prior knowledge in regression analysis. *Biometrika, 51*(1/2), 219-230.

Torgerson, W. S. (1958). *Theory and methods of scaling*. New York: Wiley.

Triggs, T. J. (1988). The ergonomics of decision-making in large scale systems: Information displays and expert knowledge elicitation. *Ergonomics, 31*, 711.

Tse, P., & Atherton, D. (1999). Prediction of machine deterioration using vibration based fault trends and recurrent neural networks. *Transactions of the ASME: Journal of Vibration and Acoustics, 121*(3), 355-362.

Tversky, A. (1974). Assessing uncertainty. *Journal of the Royal Statistical Society, Series B: Methodological, 36*(2), 148-159.

Tversky, A., & Kahneman, D. (1973). Availability: A heuristic for judging frequency and probability. *Cognitive Psychology, 5*(2), 207-232.

Tversky, A., & Kahneman, D. (1971). Belief in the law of small numbers. *Psychological Bulletin, 76*(2), 105-110.

Tversky, A., & Koehler, D. J. (1994). Support theory: A nonextensional representation of subjective probability. *Psychological Review, 101*(4), 547-567.

Uebele, V. A., & Lan, S. M. S. (1995). A neural-network-based fuzzy classifier. *IEEE Transactions on Systems, Man and Cybernetics, 25*(2), 353-361.

Vlok, P. J., Coetzee, J. L., Banjevic, D., Jardine, A. K. S., & Makis, V. (2002). Optimal component replacement decisions using vibration monitoring and the proportional-hazards model. *Operations Research Society, 53*, 193-202.

Wach, D. (2003). On-line monitoring and dynamic feature trending as a means to improve in-service inspection, maintenance, and long-term assessment of systems and components. *Nuclear Technology, 141*(1), 54-62.

Wagner, W. P., Otto, J. & Chung, Q. B. (2002). *Knowledge acquisition for expert systems in accounting and financial problem domains*.http://www30.homepage.villanova.edu/q.chung/r/kbs2002.pdf

Walls, L., & Quigley, J. (2001). Building prior distributions to support Bayesian reliability growth modelling using expert judgement. *Reliability Engineering and System Safety, 74*(2), 117-128.

Wallsten, T. S., & Budescu, D. V. (1983). Encoding subjective probabilities: A psychological and psychometric review. *Management Science, 29*(2), 151-173.

Wallsten, T. S., Budescu, D. V., & Zwick, R. (1993). Comparing the calibration and coherence of numerical and verbal probability judgments. *Management Science, 39*(2), 176-190.

Wegerich, S., & Wilks, A. (2000). *Functionality of the smart signal modeling and detection engine*. Illinois: Smart Signal Corporation.

Welbank, M. (1990). An overview of knowledge acquisition methods. *Interacting with Computers, 2*(1), 83-91.

Westerkamp, T. A. (1997). *Maintenance manager's standard manual*. Prentice Hall.

Williams, J. H., Davies, A., & Drake, P. R. (1994). *Condition-based maintenance and machine diagnostics*. Chapman & Hall.

Wilson, M. (1989). Task models for knowledge elicitations. In D. Diaper (Ed.), *Knowledge elicitation : Principles, techniques, and applications*. New York: Halsted Press.

Winkler, R. L. (1980). Prior information, predictive distributions, and Bayesian model-building. In A.Zellner (Ed.), *Bayesian analysis in econometrics and statistics*. Amsterdam: North Holland.

Winkler, R. L. (1972). *Introduction to Bayesian inference and decision*. Holt, Rinehart and Winston.

Winkler, R. L. (1967). The assessment of prior distributions in Bayesian analysis. *Journal of the American Statistical Association*, *62*(319), 776-800.

Wireman, T. (2005). *Developing performance indicators for managing maintenance*. New York: Industrial Press.

Wireman, T. (2003). *Benchmarking best practices in maintenance management*. New York: Industrial Press.

Wireman, T. (1994). *Computerized maintenance management* (2nd ed.). New York: Industrial Press.

Wireman, T. (1990). *World class maintenance management*. New York: Industrial Press.

Young, F. W. *Multidimensional scaling*.http://ccrma-www.stanford.edu/~unjung/mylec/mds.html

Zhang, C., Le, M. T., Seth, B. B., & Liang, S. Y. (2002). Bearing life prognosis under environmental effects based on accelerated life testing. *Proceedings of the Institution of Mechanical Engineers, Part C: Journal of Mechanical Engineering Science*, *216*(5), 509-516.

Zuashkiani, A., Banjevic, D., & Jardine, Andrew K. S. (2006). Incorporating expert knowledge when estimating parameters of the proportional hazards model. *Annual Reliability and Maintainability Symposium*, Newport Beach, CA.

Zuashkiani, A., Banjevic, D., & Jardine, Andrew K. S. (2004). Condition-based maintenance techniques. *CORSE-INFORMS Conference*, Banff, AB.

Zyl, R. V. (2002). *Optimization of bellis and morcom 3rd-stage piston ring CBM model*. Final EXAKT evaluation report. Hamilton: Dofasco Utilities.

14. APPENDICES

Appendix A

List of case comparisons and case analyses questions for the case study in Dofasco Inc.

Case Comparison Questions

Type 1 Question

Please compare cases A and B in each table in terms of probability of having a failure (risk of failure). Please also assume that other components such as valves and intercooler system are working properly and we only consider failures of the 2nd-stage piston ring. Please try to follow terminology shown below when comparing the two cases:

- Case A (B) has much higher risk of failure compared to Case B (A)
- Case A (B) has higher risk of failure compared to Case B (A)
- Case A (B) has slightly higher risk of failure compared to Case B (A)
- Case A (B) has almost the same risk of failure compared to Case B (A).
- These cases do not make any sense based on my experience
- Or any other explanation that you think is appropriate which I have not mentioned here.

Case 1:

	Case A	Case B
2nd-stage discharge gas temperature (F)	320	310
3rd-stage discharge gas temperature (F)	325	330
2nd-stage discharge gas pressure (psi)	140	140
Age (months)	4	4

Answer: Case A and Case B have the same risk.

Case 2:

	Case A	Case B
2nd-stage discharge gas temperature (F)	300	310
3rd-stage discharge gas temperature (F)	325	330
2nd-stage discharge gas pressure (psi)	150	140
Age (months)	4	4

Answer: Case A has slightly higher risk of failure compared to Case B.

Case 3:

	Case A	Case B
2nd-stage discharge gas temperature (F)	310	310
3rd-stage discharge gas temperature (F)	320	330
2nd-stage discharge gas pressure (psi)	150	140
Age (months)	4	4

Answer: It does not make sense.

Case 4:

	Case A	Case B
2nd-stage discharge gas temperature (F)	310	300
3rd-stage discharge gas temperature (F)	320	330
2nd-stage discharge gas pressure (psi)	140	140
Age (months)	4	4

Answer: Case B has higher risk of failure compared to Case A.

Case5:

	Case A	Case B
2nd-stage discharge gas temperature (F)	310	300
3rd-stage discharge gas temperature (F)	325	325
2nd-stage discharge gas pressure (psi)	140	150
Age (months)	4	4

Answer: Case B has much higher risk of failure compared to Case A.

Case 6:

	Case A	Case B
2nd-stage discharge gas temperature (F)	280	300
3rd-stage discharge gas temperature (F)	320	330
2nd-stage discharge gas pressure (psi)	140	140
Age (months)	5	3

Answer: It does not make sense.

Type 2 Question

Please compare cases A and B in each table in terms of probability of having a failure (risk of failure). Also assume that other components such as valves and intercooler system are working properly and we only consider failures of the 2nd-stage piston ring. However what makes this case comparisons different from previous case comparisons is that it is obvious that case B is at higher risk of failure (since the values of the condition indicators in case B are more than or equal to those of case A).The question is that how many times case B is at more risk of failure compared to case A. For example you can say that case B is between 2-4 times riskier compared to A. Use the upper and lower limits which make you 90% confident that the right value would be between them. For example when you say case B is between 2 and 4 times riskier than case A this should also imply that you are 90% sure of your answer. If you are not sure you can expand the interval to a point that makes you confident. For instance in this case if you think interval between 2 and 4 does not make you 90% confident, you can use 1.5 and 4.5 as lower and upper limits.

Question Set One

Case 1:

	Case A	**Case B**
2nd-stage discharge gas temperature (F)	300	310
3rd-stage discharge gas temperature (F)	325	330
2nd-stage discharge gas pressure (psi)	140	140
Age (months)	4	4

Answer: B is between 1.1 and 1.2 times riskier than A.

Temperatures are only slightly higher. Pressure/age are all the same. Only a slightly higher risk based on 3rd stg temperature.

Case 2:

	Case A	**Case B**
2nd-stage discharge gas temperature (F)	300	310
3rd-stage discharge gas temperature (F)	320	330
2nd-stage discharge gas pressure (psi)	140	140
Age (months)	4	4

Answer: B is between 1.2 and 1.5 times riskier than A.

More gap in temperatures than Case 1. 2nd stg pressures are still equal though and we consider 140 psi to be normal.

Case 3:

	Case A	Case B
2nd-stage discharge gas temperature (F)	300	310
3rd-stage discharge gas temperature (F)	320	330
2nd-stage discharge gas pressure (psi)	135	140
Age (months)	4	4

Answer: B is between 2 and 3 times riskier than A.

All key indicators in Case B are higher than Case A. 2nd stg pressure is still normal though. 3rd stg temperature is quite high. Definitely higher risk of failure.

Case 4:

	Case A	Case B
2nd-stage discharge gas temperature (F)	305	310
3rd-stage discharge gas temperature (F)	325	330
2nd-stage discharge gas pressure (psi)	140	145
Age (months)	4	4

Answer: B is between 2 and 3.5 times riskier than A.

Very similar to Case 3. I would predict even a higher risk of failure except the temperatures in Case 4 are a little closer together. 2nd stg pressure is considered high.

Case 5:

	Case A	Case B
2nd-stage discharge gas temperature (F)	310	310
3rd-stage discharge gas temperature (F)	325	330
2nd-stage discharge gas pressure (psi)	140	140
Age (months)	4	4

Answer: B is between 1.1 and 1.2 times riskier than A.

Very similar to Case 1. only a slightly higher risk based on 3rd stg temperature.

Case 6:

	Case A	Case B
2nd-stage discharge gas temperature (F)	305	310
3rd-stage discharge gas temperature (F)	325	325
2nd-stage discharge gas pressure (psi)	140	140
Age (months)	4	4

Answer: No higher risk.

Case 7:

	Case A	Case B
2nd-stage discharge gas temperature (F)	310	310
3rd-stage discharge gas temperature (F)	330	330
2nd-stage discharge gas pressure (psi)	145	150
Age (months)	4	4

Answer: B is between 1.5 and 2 times riskier than A.

Temperatures are the same but the 2nd stg pressure on B is very high. However, 2nd stg pressure in A is also high.

Case 8:

	Case A	Case B
2nd-stage discharge gas temperature (F)	300	310
3rd-stage discharge gas temperature (F)	320	320
2nd-stage discharge gas pressure (psi)	140	145
Age (months)	4	4

Answer: B is between 1.5 and 2 times riskier than A.

This is a tough call. 2nd stg pressure is high. Temperatures are normal though.

Case 9:

	Case A	Case B
2nd-stage discharge gas temperature (F)	310	310
3rd-stage discharge gas temperature (F)	325	330
2nd-stage discharge gas pressure (psi)	140	145
Age (months)	4	4

Answer: B is between 1.5 and 2 times riskier than A.

Case 10:

	Case A	Case B
2nd-stage discharge gas temperature (F)	300	310
3rd-stage discharge gas temperature (F)	325	325
2nd-stage discharge gas pressure (psi)	140	145
Age (months)	4	4

Answer: B is between 1.5 and 1.9 times riskier than A.

Not as much risk as Case 9, only because 3rd stg temperatures are still the same.

Case 11:

	Case A	Case B
2nd-stage discharge gas temperature (F)	305	310
3rd-stage discharge gas temperature (F)	325	330
2nd-stage discharge gas pressure (psi)	140	145
Age (months)	4	4

Answer: B is between 2 and 3.5 times riskier than A.

Same as Case 4

Case 12:

	Case A	Case B
2nd-stage discharge gas temperature (F)	300	300
3rd-stage discharge gas temperature (F)	320	320
2nd-stage discharge gas pressure (psi)	135	140
Age (months)	4	4

Answer: B is between 1.1 and 1.5 times riskier than A.

Case 13:

	Case A	Case B
2nd-stage discharge gas temperature (F)	300	305
3rd-stage discharge gas temperature (F)	320	325
2nd-stage discharge gas pressure (psi)	135	140
Age (months)	4	4

Answer: B is between 2 and 3 times riskier than A.

Case 14:

	Case A	Case B
2nd-stage discharge gas temperature (F)	300	305
3rd-stage discharge gas temperature (F)	315	320
2nd-stage discharge gas pressure (psi)	135	140
Age (months)	4	4

Answer: B is between 2 and 3 times riskier than A.

Case 15:

	Case A	Case B
2nd-stage discharge gas temperature (F)	305	310
3rd-stage discharge gas temperature (F)	330	330
2nd-stage discharge gas pressure (psi)	145	150
Age (months)	4	4

Answer: B is between 1.5 and 2 times riskier than A.

Question Set Two

Case 1:

	Case A	Case B
2nd-stage discharge gas temperature (F)	300	310
3rd-stage discharge gas temperature (F)	325	330
2nd-stage discharge gas pressure (psi)	143	143
Age (months)	4	4

Answer: B is between 1 and 2 times riskier than A.

Based on slightly elevated 2nd and 3rd stg temperatures. 2nd stg pressure is elevated in both cases.

Case 2:

	Case A	Case B
2nd-stage discharge gas temperature (F)	300	310
3rd-stage discharge gas temperature (F)	320	330
2nd-stage discharge gas pressure (psi)	143	143
Age (months)	4	4

Answer: B is between 1.5 and 2.5 times riskier than A.

Case 3:

	Case A	Case B
2nd-stage discharge gas temperature (F)	300	310
3rd-stage discharge gas temperature (F)	320	330
2nd-stage discharge gas pressure (psi)	140	145
Age (months)	4	4

Answer: B is between 2 and 4 times riskier than A.

All indicators are elevated.

Case 4:

	Case A	Case B
2nd-stage discharge gas temperature (F)	310	310
3rd-stage discharge gas temperature (F)	325	330
2nd-stage discharge gas pressure (psi)	143	143
Age (months)	4	4

Answer: B is between 1 and 1.5 times riskier than A.

Very close. Slightly higher only because of 3rd stg temperature.

Case 5:

	Case A	Case B
2nd-stage discharge gas temperature (F)	305	310
3rd-stage discharge gas temperature (F)	325	325
2nd-stage discharge gas pressure (psi)	143	143
Age (months)	4	4

Answer: No difference. 2nd stg pressure and 3rd stg temperature are equal.

Case 6:

	Case A	Case B
2nd-stage discharge gas temperature (F)	305	310
3rd-stage discharge gas temperature (F)	330	330
2nd-stage discharge gas pressure (psi)	145	145
Age (months)	4	4

Answer: No difference.

Case 7:

	Case A	Case B
2nd-stage discharge gas temperature (F)	310	310
3rd-stage discharge gas temperature (F)	330	330
2nd-stage discharge gas pressure (psi)	142	147
Age (months)	4	4

Answer: B is between 1.5 and 2.5 times riskier than A.

Temperatures are the same in both cases but 2nd stg pressure on Case B higher probability of failure.

Case 8:

	Case A	Case B
2nd-stage discharge gas temperature (F)	310	310
3rd-stage discharge gas temperature (F)	330	330
2nd-stage discharge gas pressure (psi)	142	145
Age (months)	4	4

Answer: B is between 1 and 2 times riskier than A.

Case 9:

	Case A	Case B
2nd-stage discharge gas temperature (F)	300	310
3rd-stage discharge gas temperature (F)	325	325
2nd-stage discharge gas pressure (psi)	140	145
Age (months)	4	4

Answer: B is between 1.5 and 3 times riskier than A.

Case 10:

	Case A	Case B
2nd-stage discharge gas temperature (F)	305	310
3rd-stage discharge gas temperature (F)	325	325
2nd-stage discharge gas pressure (psi)	142	142
Age (months)	4	4

Answer: No difference

Case 11:

	Case A	Case B
2nd-stage discharge gas temperature (F)	310	310
3rd-stage discharge gas temperature (F)	325	330
2nd-stage discharge gas pressure (psi)	145	145
Age (months)	4	4

Answer: B is between 1.1 and 1.5 times riskier than A.

Slightly higher only on 3rd stg temperature.

Case 12:

	Case A	Case B
2nd-stage discharge gas temperature (F)	305	310
3rd-stage discharge gas temperature (F)	327	327
2nd-stage discharge gas pressure (psi)	140	145
Age (months)	4	4

Answer: B is between 1.5 and 3 times riskier than A.

Case 13:

	Case A	Case B
2nd-stage discharge gas temperature (F)	300	300
3rd-stage discharge gas temperature (F)	320	320
2nd-stage discharge gas pressure (psi)	140	140
Age (months)	3.5	4

Answer: No difference.

Case 14:

	Case A	Case B
2nd-stage discharge gas temperature (F)	300	300
3rd-stage discharge gas temperature (F)	320	320
2nd-stage discharge gas pressure (psi)	140	140
Age (months)	3	4

Answer: B is between 1 and 3 times riskier than A.

Case B will start to trend up in temperatures and pressures as it approaches the 5 month ring life expectancy.

Case 15:

	Case A	Case B
2nd-stage discharge gas temperature (F)	300	300
3rd-stage discharge gas temperature (F)	320	320
2nd-stage discharge gas pressure (psi)	140	140
Age (months)	4	4.5

Answer: B is between 2 and 4 times riskier than A.

Approaching ring life expectancy. Pressure and temperatures will soon start to rise and give early warning that rings require replacing.

Case 16:

	Case A	Case B
2nd-stage discharge gas temperature (F)	310	310
3rd-stage discharge gas temperature (F)	325	325
2nd-stage discharge gas pressure (psi)	142	142
Age (months)	3.5	4

Answer: Case A is probably at a higher risk of failure as the 2nd stg pressure is in alarm condition even though the rings are younger in Age.

Case 17:

	Case A	Case B
2nd-stage discharge gas temperature (F)	310	310
3rd-stage discharge gas temperature (F)	325	325
2nd-stage discharge gas pressure (psi)	142	142
Age (months)	4	4.5

Answer: Similar to Case 16 above. There is a definite relationship which is not shown in these two cases. 2nd stg pressure and temperatures should be lower in Case A because they have not been in service long.

Case 18:

	Case A	Case B
2nd-stage discharge gas temperature (F)	310	310
3rd-stage discharge gas temperature (F)	325	325
2nd-stage discharge gas pressure (psi)	142	142
Age (months)	4.5	5

Answer: B is between 1 and 2 times riskier than A.

Both Case A and B are showing signs of failure. Case B has slightly higher risk of failure based on age.

Case 19:

	Case A	Case B
2nd-stage discharge gas temperature (F)	310	310
3rd-stage discharge gas temperature (F)	330	330
2nd-stage discharge gas pressure (psi)	145	145
Age (months)	4.5	5

Answer: B is between 1 and 2 times riskier than A.

Very similar to Case 18. I would say Case B is slightly higher based on Age.

Case 20:

	Case A	Case B
2nd-stage discharge gas temperature (F)	310	310
3rd-stage discharge gas temperature (F)	330	330
2nd-stage discharge gas pressure (psi)	145	145
Age (months)	4	5

Answer: B is between 1 and 3 times riskier than A.

This is tough. Numbers tell me that the rings in Case A are in trouble and they are younger than Case B, however, presented with these two sets of identical numbers, I will still go with Case B based on Age

Questions That Define Upper and Lower Bounds of the Hazard

Consider failure of 3rd-stage piston rings. Please assume that other components such as valves, intercooler system and etc are working properly and we only consider failures of the 2nd-stage piston ring and the component has not failed yet. For each case please give your best estimate of probability of having a failure during the next 24 hours and also the average of remaining time to failure. You can provide intervals for the required answers, for example you can say: "I am 90% confident that the item will fail in 2-5 days."

Cases are described below:

Case 1

	Sample Case
2nd-stage discharge gas temperature (F)	310
3rd-stage discharge gas temperature (F)	330
2nd-stage discharge gas pressure (psi)	145
Age (months)	4.5

Probability of having a failure within the next 24 hours from now: **70**

Mean time to failure from now (how long does it take in average for the component to fail from now): **2-3**

Case 2

	Sample Case
2nd-stage discharge gas temperature (F)	310
3rd-stage discharge gas temperature (F)	330
2nd-stage discharge gas pressure (psi)	142
Age (months)	4.5

Probability of having a failure within the next 24 hours from now: **60**

Mean time to failure from now: **3-5**

Case 3

	Sample Case
2nd-stage discharge gas temperature (F)	310
3rd-stage discharge gas temperature (F)	320
2nd-stage discharge gas pressure (psi)	145
Age (months)	4.5

Probability of having a failure within the next 24 hours from now: **50**

Mean time to failure from now: **2-5**

Case 4

	Sample Case
2nd-stage discharge gas temperature (F)	300
3rd-stage discharge gas temperature (F)	330
2nd-stage discharge gas pressure (psi)	145
Age (months)	4.5

Probability of having a failure within the next 24 hours from now: **70**

Mean time to failure from now: **2-3**

Case 5

	Sample Case
2nd-stage discharge gas temperature (F)	310
3rd-stage discharge gas temperature (F)	330
2nd-stage discharge gas pressure (psi)	140
Age (months)	4.5

Probability of having a failure within the next 24 hours from now: **50**

Mean time to failure from now: **3-5**

Case 6

	Sample Case
2nd-stage discharge gas temperature (F)	300
3rd-stage discharge gas temperature (F)	310
2nd-stage discharge gas pressure (psi)	145
Age (months)	4.5

Probability of having a failure within the next 24 hours from now: **50**

Mean time to failure from now: **2-5**

Case 7

	Sample Case
2nd-stage discharge gas temperature (F)	300
3rd-stage discharge gas temperature (F)	305
2nd-stage discharge gas pressure (psi)	140
Age (months)	4.5

Probability of having a failure within the next 24 hours from now: **0**

Mean time to failure from now: **10-14**

Case 8

	Sample Case
2nd-stage discharge gas temperature (F)	310
3rd-stage discharge gas temperature (F)	330
2nd-stage discharge gas pressure (psi)	145
Age (months)	4

Probability of having a failure within the next 24 hours from now: **75**

Mean time to failure from now: **2-3**

Case 9

	Sample Case
2nd-stage discharge gas temperature (F)	310
3rd-stage discharge gas temperature (F)	330
2nd-stage discharge gas pressure (psi)	155
Age (months)	4.5

Probability of having a failure within the next 24 hours from now: **100**

Mean time to failure from now: **1**

Case 10

	Sample Case
2nd-stage discharge gas temperature (F)	300
3rd-stage discharge gas temperature (F)	325
2nd-stage discharge gas pressure (psi)	142
Age (months)	4.5

Probability of having a failure within the next 24 hours from now: **50**

Mean time to failure from now: **3-5**

Case 11

	Sample Case
2nd-stage discharge gas temperature (F)	300
3rd-stage discharge gas temperature (F)	310
2nd-stage discharge gas pressure (psi)	140
Age (months)	4.5

Probability of having a failure within the next 24 hours from now: **0**

Mean time to failure from now: **10-14**

Case 12

	Sample Case
2nd-stage discharge gas temperature (F)	290
3rd-stage discharge gas temperature (F)	305
2nd-stage discharge gas pressure (psi)	140
Age (months)	4.5

Probability of having a failure within the next 24 hours from now: **0**

Mean time to failure from now: **10-14**

Case 13

	Sample Case
2nd-stage discharge gas temperature (F)	295
3rd-stage discharge gas temperature (F)	305
2nd-stage discharge gas pressure (psi)	142
Age (months)	4.5

Probability of having a failure within the next 24 hours from now: **0**

Mean time to failure from now: **7-10**

APPENDIX B

Parameters estimated based on statistical data and on the combination of expert knowledge and statistical data

	E + D0	E + D1	E + D2	E + D3	E + D4	E + D5	E + D6	E + D7	E + D8	E + D9
β	1.5593	1.5595	1.5618	1.5625	1.5630	1.5650	1.5624	1.5543	1.5539	1.5544
A	41.1004	41.1217	41.4799	41.5299	41.4857	41.6172	41.3921	41.0665	41.0087	41.0240
γ_1	0.1214	0.1213	0.1206	0.1207	0.1209	0.1207	0.1210	0.1214	0.1215	0.1215
γ_2	0.0070	0.0070	0.0070	0.0070	0.0070	0.0070	0.0069	0.0069	0.0069	0.0069
γ_3	0.0540	0.0540	0.0550	0.0550	0.0548	0.0552	0.0546	0.0539	0.0538	0.0539
Median of measure of accuracy	1.8054	1.7881	1.4858	1.4641	1.5131	1.4363	1.5658	1.7554	1.8549	1.8854
Correlation of h_{E+Dj} and h_{D39}	0.9883	0.9883	0.9889	0.9888	0.9887	0.9889	0.9886	0.9882	0.9882	0.9882

	E + D10	E + D11	E + D12	E + D13	E + D14	E + D15	E + D16	E + D17	E + D18	E + D19
β	1.5562	1.5560	1.5562	1.5557	1.5585	1.5634	1.5666	1.5687	1.5692	1.5712
A	41.0780	41.2890	41.2938	41.2864	41.3816	41.5508	41.6062	41.6844	41.6825	41.7186
γ_1	0.1217	0.1215	0.1215	0.1208	0.1209	0.1209	0.1207	0.1205	0.1205	0.1203
γ_2	0.0069	0.0069	0.0069	0.0069	0.0069	0.0070	0.0070	0.0070	0.0070	0.0070
γ_3	0.0539	0.0543	0.0543	0.0548	0.0548	0.0548	0.0548	0.0552	0.0552	0.0552
Median of measure of accuracy	1.8573	1.6304	1.6273	1.7198	1.5771	1.3895	1.3300	1.3127	1.3216	1.2939
Correlation of h_{E+Dj} and h_{D39}	0.9881	0.9883	0.9883	0.9888	0.9887	0.9887	0.9888	0.9890	0.9890	0.9891

	E + D20	E + D21	E + D22	E + D23	E + D24	E + D25	E + D26	E + D27	E + D28	E + D29
β	1.5727	1.5737	1.5712	1.5714	1.5729	1.5711	1.5718	1.5728	1.5738	1.5705
A	41.7726	41.7917	41.6712	41.6755	41.7886	41.7145	41.7843	41.8328	41.8167	41.8082
γ_1	0.1201	0.1201	0.1200	0.1200	0.1196	0.1197	0.1196	0.1195	0.1197	0.1194
γ_2	0.0070	0.0070	0.0070	0.0070	0.0070	0.0070	0.0070	0.0070	0.0070	0.0070
γ_3	0.0555	0.0555	0.0554	0.0553	0.0557	0.0555	0.0558	0.0559	0.0557	0.0560
Median of measure of accuracy	1.2863	1.2795	1.3217	1.3014	1.2562	1.2865	1.2648	1.2490	1.2417	1.2582
Correlation of h_{E+Dj} and h_{D39}	0.9892	0.9892	0.9893	0.9892	0.9895	0.9894	0.9895	0.9896	0.9894	0.9897

	E + D30	E + D31	E + D32	E + D33	E + D34	E + D35	E + D36	E + D37	E + D38	E + D39
β	1.5708	1.5718	1.5732	1.5738	1.5748	1.5738	1.5745	1.5758	1.5766	1.5778
A	41.7532	41.7580	41.7502	41.8548	41.8855	41.9475	41.9245	41.9631	42.0033	42.0972
γ_1	0.1192	0.1196	0.1196	0.1195	0.1195	0.1194	0.1195	0.1193	0.1191	0.1190
γ_2	0.0069	0.0069	0.0069	0.0070	0.0070	0.0070	0.0070	0.0070	0.0070	0.0070
γ_3	0.0560	0.0558	0.0557	0.0560	0.0561	0.0563	0.0561	0.0563	0.0564	0.0568
Median of measure of accuracy	1.2799	1.2845	1.2891	1.2485	1.2387	1.2170	1.2140	1.2016	1.1920	1.1591
Correlation of h_{E+Dj} and h_{D39}	0.9898	0.9895	0.9895	0.9896	0.9897	0.9898	0.9896	0.9898	0.9899	0.9901

	D0	D1	D2	D3	D4	D5	D6	D7	D8	D9
β	N/A*	N/A	N/A	3.1160	3.6083	3.9140	2.6298	0.9508	0.8415	0.8348
A	N/A	N/A	N/A	57.5398	52.5989	54.9054	40.2025	35.4969	44.1785	49.1279
γ_1	N/A	N/A	N/A	0.1705	0.2081	0.2194	0.1774	0.1166	0.1207	0.1270
γ_2	N/A	N/A	N/A	0.0651	0.0226	0.0202	0.0000	0.0000	0.0000	0.0000
γ_3	N/A	N/A	N/A	0.0000	0.0006	0.0000	0.0129	0.0533	0.0830	0.0968
Median of measure of accuracy	N/A	N/A	N/A	0.1122	0.2456	0.1887	1.0173	2.5329	2.1549	1.8058
Correlation of h_{Dj} and h_{D39}	N/A	N/A	N/A	0.9397	0.9155	0.9101	0.9341	0.9917	0.9947	0.9919

	D10	D11	D12	D13	D14	D15	D16	D17	D18	D19
β	1.2420	1.0235	1.1011	0.9660	1.4289	1.7471	1.8704	1.9512	2.0070	2.0513
A	45.8648	48.7754	48.5652	55.9914	44.3917	41.0697	38.5534	40.5420	41.6927	40.2323
γ_1	0.1418	0.1481	0.1497	0.1212	0.1344	0.1410	0.1332	0.1299	0.1345	0.1235
γ_2	0.0000	0.0000	0.0000	0.0000	0.0000	0.0000	0.0000	0.0000	0.0000	0.0000
γ_3	0.0725	0.0822	0.0795	0.1204	0.0671	0.0474	0.0406	0.0476	0.0485	0.0481
Median of measure of accuracy	1.9176	1.5546	1.5486	1.2085	1.4681	1.3765	1.4396	1.2909	1.2712	1.2914
Correlation of h_{Dj} and h_{D39}	0.9854	0.9842	0.9832	0.9884	0.9873	0.9773	0.9773	0.9823	0.9804	0.9853

	D20	D21	D22	D23	D24	D25	D26	D27	D28	D29
β	2.0301	2.0561	1.9300	1.8731	1.8454	1.8233	1.8155	1.8375	1.8555	1.7687
A	41.9716	41.9935	38.0415	36.4534	37.0384	36.4514	37.5861	38.1009	37.9928	37.4165
γ_1	0.1116	0.1122	0.0795	0.0755	0.0630	0.0663	0.0652	0.0666	0.0787	0.0627
γ_2	0.0012	0.0003	0.0003	0.0003	0.0063	0.0030	0.0053	0.0051	0.0087	0.0031
γ_3	0.0586	0.0589	0.0628	0.0600	0.0625	0.0625	0.0647	0.0655	0.0560	0.0680
Median of measure of accuracy	1.1505	1.1373	1.2576	1.2562	1.1777	1.2537	1.1796	1.1316	1.2102	1.1197
Correlation of h_{Dj} and h_{D39}	0.9937	0.9937	0.9989	0.9977	0.9915	0.9928	0.9924	0.9932	0.9993	0.9875

	D30	D31	D32	D33	D34	D35	D36	D37	D38	D39
β	1.8071	1.8642	1.8998	1.8697	1.8855	1.8159	1.8370	1.8606	1.8674	1.8614
A	35.8978	38.4560	38.3788	39.3641	39.6761	40.8280	39.9219	39.5205	39.8707	40.3916
γ_1	0.0483	0.0802	0.0813	0.0837	0.0846	0.0878	0.0934	0.0900	0.0890	0.0874
γ_2	0.0000	0.0000	0.0000	0.0021	0.0023	0.0046	0.0031	0.0023	0.0025	0.0032
γ_3	0.0719	0.0649	0.0637	0.0642	0.0644	0.0656	0.0610	0.0616	0.0629	0.0648
Median of measure of accuracy	1.0770	1.1456	1.1807	1.1140	1.0942	1.0471	1.0910	1.0898	1.0620	1.0000
Correlation of h_{Dj} and h_{D39}	0.9559	0.9987	0.9991	0.9998	0.9999	1.0000	0.9993	0.9998	0.9999	1.0000

The results based on the second attempt.

	E + D0	**E + D1**	**E + D2**	**E + D3**	**E + D4**	**E + D5**	**E + D6**	**E + D7**	**E + D8**	**E + D9**
β	1.5568	1.5569	1.5583	1.5582	1.5585	1.5599	1.5575	1.5501	1.5493	1.5494
A	41.5107	41.5225	41.6849	41.6633	41.5722	41.6889	41.4278	41.0584	40.9566	40.9303
γ_1	0.1215	0.1215	0.1210	0.1213	0.1217	0.1214	0.1219	0.1225	0.1229	0.1230
γ_2	0.0070	0.0070	0.0070	0.0070	0.0070	0.0070	0.0070	0.0070	0.0070	0.0070
γ_3	0.0539	0.0539	0.0545	0.0543	0.0539	0.0543	0.0535	0.0524	0.0521	0.0520
Median of measure of accuracy	1.1842	1.1792	1.1053	1.1228	1.1735	1.1270	1.2336	1.3602	1.4297	1.4664
Correlation of h_{E+Dj} and h_{D39}	0.9882	0.9882	0.9886	0.9884	0.9881	0.9883	0.9879	0.9873	0.9870	0.9869

The results based on the second attempt.

	E + D10	E + D11	E + D12	E + D13	E + D14	E + D15	E + D16	E + D17	E + D18	E + D19
β	1.5508	1.5512	1.5513	1.5507	1.5529	1.5568	1.5589	1.5606	1.5610	1.5624
A	40.9156	41.1693	41.1741	41.1532	41.2517	41.4205	41.4428	41.5150	41.5022	41.5225
γ_1	0.1233	0.1228	0.1227	0.1222	0.1220	0.1218	0.1217	0.1215	0.1215	0.1214
γ_2	0.0070	0.0070	0.0070	0.0069	0.0069	0.0070	0.0070	0.0070	0.0070	0.0070
γ_3	0.0518	0.0526	0.0527	0.0530	0.0533	0.0535	0.0535	0.0538	0.0538	0.0538
Median of measure of accuracy	1.4775	1.3595	1.3576	1.4139	1.3546	1.2324	1.1904	1.1762	1.1928	1.1725
Correlation of h_{E+Dj} and h_{D39}	0.9867	0.9872	0.9873	0.9876	0.9878	0.9879	0.9879	0.9882	0.9881	0.9882

The results based on the second attempt.

	E + D20	E + D21	E + D22	E + D23	E + D24	E + D25	E + D26	E + D27	E + D28	E + D29
β	1.5637	1.5644	1.5626	1.5627	1.5637	1.5624	1.5629	1.5636	1.5642	1.5621
A	41.5853	41.6024	41.4793	41.4597	41.5701	41.4940	41.5651	41.6116	41.5671	41.5898
γ_1	0.1211	0.1211	0.1209	0.1209	0.1205	0.1207	0.1205	0.1204	0.1207	0.1202
γ_2	0.0070	0.0070	0.0070	0.0070	0.0070	0.0070	0.0070	0.0070	0.0070	0.0069
γ_3	0.0542	0.0542	0.0541	0.0539	0.0543	0.0542	0.0544	0.0545	0.0542	0.0547
Median of measure of accuracy	1.1674	1.1600	1.2116	1.1995	1.1475	1.1869	1.1614	1.1432	1.1429	1.1626
Correlation of h_{E+Dj} and h_{D39}	0.9884	0.9885	0.9885	0.9884	0.9887	0.9886	0.9888	0.9888	0.9886	0.9890

The results based on the second attempt.

	E + D30	E + D31	E + D32	E + D33	E + D34	E + D35	E + D36	E + D37	E + D38	E + D39
β	1.5626	1.5634	1.5646	1.5648	1.5655	1.5645	1.5647	1.5658	1.5664	1.5668
A	41.5373	41.5272	41.5139	41.6111	41.6395	41.7001	41.6479	41.6816	41.7249	41.8157
γ_1	0.1200	0.1205	0.1205	0.1204	0.1204	0.1203	0.1206	0.1203	0.1202	0.1200
γ_2	0.0069	0.0069	0.0069	0.0069	0.0069	0.0070	0.0070	0.0069	0.0069	0.0069
γ_3	0.0547	0.0544	0.0544	0.0546	0.0547	0.0549	0.0546	0.0548	0.0549	0.0552
Median of measure of accuracy	1.1911	1.2024	1.2115	1.1649	1.1536	1.1276	1.1306	1.1179	1.1061	1.0726
Correlation of h_{E+Dj} and h_{D39}	0.9891	0.9888	0.9888	0.9889	0.9889	0.9890	0.9888	0.9890	0.9891	0.9892

CPSIA information can be obtained at www.ICGtesting.com
Printed in the USA
LVOW06s0318111115

461999LV00029B/259/P

9 783639 020564